RILL EROSION

Rorke B. Bryan (Editor)

RILL EROSION

Processes and Significance

CATENA SUPPLEMENT 8

CATENA – A cooperating Journal of the International Society of Soil Science

ISSS - AISS - IBG

Cover photo by Rorke B. Bryan:

Dense rills on the walls of a rapidly-expanding gully system, Karatu-Oldeani, Tanzania. These rills result from the gully development which was initiated by piping in highly erodible volcanic soils.

CIP-Kurztitelaufnahme der Deutschen Bibliothek

Rill Erosion: Processes and Significance / R.B. Bryan (ed.). –
Cremlingen-Destedt: CATENA VERLAG, 1987
(CATENA: Supplement; 8) ISBN 3-923381-07-7. NE: Bryan, Rorke B. [Hrsg.]; CATENA/Supplement

© Copyright 1987 by CATENA VERLAG, D-3302 Cremlingen 4, W. Germany

All rights are reserved. No part of this publication may be reproduced, stored in a retrieval system or transmitted in any form or by any means, electronic, mechanical, photocopying, recording or otherwise without prior permission of the publisher. Submission of an article implies the transfer of the copyright from the authors to the publisher.

This publication has been registered with the Copyright Clearance Center, Inc. Consent is given for copying of articles for personal or internal use, for the specific clients. This consent is given on the condition that the copier pay through the Center the per-copy fee for copying beyond that permitted by Sections 107 or 108 of the U.S. Copyright Law. The per-copy fee is stated in the code-line at the bottom of the first page of each article. The appropriate fee, together with a copy of the first page of the article, should be forwarded to the Copyright Clearance Center, Inc., 27 Congress Street, Salem, MA01970, U.S.A. If no code-line appears, broad consent to copy has not been given and permission to copy must be obtained directly from the publisher. This consent does not extend to other kinds of copying, such as for general distribution, resale, advertising and promotion purposes, or for creating new collective works. Special written permission must be obtained from the publisher for such copying.

ISSN 0722–0723 / ISBN 3-923381-07-7

CONTENTS

R.B. BRYAN
PROCESSES AND SIGNIFICANCE OF RILL DEVELOPMENT   1

G. GOVERS
SPATIAL AND TEMPORAL VARIABILITY IN RILL DEVELOPMENT PROCESSES AT THE HULDENBERG EXPERIMENTAL SITE   17

J. POESEN
TRANSPORT OF ROCK FRAGMENTS BY RILL FLOW—A FIELD STUDY   35

O. PLANCHON, E. FRITSCH & C. VALENTIN
RILL DEVELOPMENT IN A WET SAVANNAH ENVIRONMENT   55

R.J. LOCH & E.C. THOMAS
RESISTANCE TO RILL EROSION: OBSERVATIONS ON THE EFFICIENCY OF RILL EROSION ON A TILLED CLAY SOIL UNDER SIMULATED RAIN AND RUN-ON WATER   71

M.A. FULLEN & A.H. REED
RILL EROSION ON ARABLE LOAMY SANDS IN THE WEST MIDLANDS OF ENGLAND   85

D. TORRI, M. SFALANGA & G. CHISCI
THRESHOLD CONDITIONS FOR INCIPIENT RILLING   97

G. RAUWS
THE INITIATION OF RILLS ON PLANE BEDS OF NON-COHESIVE SEDIMENTS   107

D.C. FORD & J. LUNDBERG
A REVIEW OF DISSOLUTIONAL RILLS IN LIMESTONE AND OTHER SOLUBLE ROCKS   119

J. GERITS, A.C. IMESON, J.M. VERSTRATEN & R.B. BRYAN
RILL DEVELOPMENT AND BADLAND REGOLITH PROPERTIES   141

PROCESSES AND SIGNIFICANCE OF RILL DEVELOPMENT

R.B. Bryan, Scarborough

Some of the confusion about the geomorphic significance of rills reflects the wide range of features to which the term has been applied by different researchers. It would be overly sanguine to hope that this collection of recent research on rilling could provide a fully integrated assessment of the role of the process in all geomorphic systems. It should, however, indicate how the wide range of studies on rills are related, and identify some of the research which will be necessary to produce such an integrated assessment. This introduction is not intended to be a comprehensive review of all studies relevant to rill development, but an attempt to show the relationship between the papers in the collection and recent major trends in research.

Most early research on rills was carried out for agricultural purposes and the most commonly-used definition still emphasizes agricultural applications— "microchannels ... small enough to be removed by normal tillage operations" (FAO 1965). This definition may explain the lack of attention paid to rill processes, for it implies relative insignificance in agriculture, but at the same time diverts attention from rill development in areas which are not repeatedly disturbed by tillage, where more permanent rills can develop which may be of greater geomorphic importance.

The potential geomorphic significance of rills as hillslope landforms, as conduits for water and sediment transport and as embryo drainage systems was first clearly recognized by HORTON (1945), who based his concept of drainage basin evolution on sequential development of rill systems by headward extension, micropiracy and cross-grading. HORTON's view of the rilling process was related to his concepts of infiltration and runoff initiation, and was derived largely from plot experiments carried out on agricultural soils. Initial particle entrainment was attributed to the shear stress exerted by runoff, as it progressively increases in depth downslope. From this came the concept of a "belt of no erosion", whose width at any location, neglecting the influence of rainsplash or other erosional processes, reflects the interaction of infiltration/runoff, slope steepness and surface resistance. It is not absolutely clear from HORTON's 1945 paper whether he regarded the "belt of no erosion" as being coincident with the absence of rilling, though he states (p. 332) that "sheet erosion implies the formation of either a rilled or a gullied surface". He attributed the concentration of erosion in channels essential to rill development to slight accidental variations in topography, which produce localized increase in runoff depth and shear stress. HORTON did not specifically discuss changes in flow hydraulics associated with rill incision, but elsewhere he did describe the mixture of laminar and turbulent flow on hillslopes, and the potential importance of rain-wave trains in initiating particle entrainment.

Further refinement of HORTON's concepts was prevented by his death, and important questions about rilling processes remained unanswered. Some were eventually addressed in several research projects, while others have been largely ignored. In any case, for almost quarter of a century after HORTON's death little research concerning rill development was reported. One major exception was SCHUMM's (1956) study of hillslope and drainage system evolution in the Perth Amboy badlands, United States. This study was primarily morphological and qualitative, and provided little precise information on rill initiation processes. However, it did validate some of HORTON's concepts and introduced new aspects of rill development. The existence of Hortonian belts of no erosion (or no rill incision) was demonstrated and, in comparatively homogeneous material, shown to depend in width primarily on slope angle. This confirmed the concept of a threshold length of flow necessary to produce the hydraulic conditions required for rill initiation, but SCHUMM expanded this concept to a threshold drainage area, and developed his concept of a "constant of channel maintenance", which addresses not only the width of the belt of no erosion, but the characteristically even spacing of rills on essentially homogeneous material.

Another important aspect of rill development which SCHUMM discussed, was their permanence. He distinguished small rills on the upper slope which developed during rainy summers and were obliterated by frost action each winter from larger, more permanent channels downslope which persisted through winter disturbance. A similar seasonal cycle of rill obliteration by frost action was described by SCHUMM & LUSBY (1963) from Mancos Shale badlands in western Colorado, while BRYAN & PRICE (1980) described seasonal rill obliteration by shallow mass movement on lake bluffs in southern Canada.

HORTON's (1945) paper did not address the question of rill permanence or identify different types of rill, yet this appears to be fundamental consideration affecting their potential function as embryo drainage systems. Permanent rills which persist in the same location for prolonged periods typically develop into gullies downslope and can form the initial stages of the evolution of major drainage systems. The potential influence of ephemeral rills which experience seasonal obliteration, is less clear. Some become deeper and wider downslope, and merge into permanent rills as described by SCHUMM (1956). Others become broader and shallower, disappearing eventually in braided washes. The influence of either type of ephemeral rill on drainage system evolution depends largely on whether the seasonal reappearance of rills always occurs in the same location. This is likely to occur if rill development is strongly constrained by topographic features, but not if rill initiation is largely a response to extreme rainstorms. HORTON recognized the close relationship between rill initiation and rainstorm characteristics, yet little subsequent work has been done on the relationship between magnitude-frequency characteristics of rainstorms and rill development at any location, or the possibility that systematic local variations in storm incidence may significantly affect the pattern of rill system development. GOVERS (this volume) does describe important variations in the character of rill processes in response to different types of rainstorm at the Huldenberg Experimental Site in Belgium. Further research on this topic is essential if the general influence of rills on drainage system evolution is to be properly assessed. It should be emphasized that, even if ephemeral rills do not always act as embryo drainage systems as envisaged by HORTON, they may still strongly affect hillslope water and sediment transport, and therefore geomorphic evolution.

The hydraulic conditions of overland flow and the resistance of surface materials are fundamental to HORTON's concepts, yet some aspects of their relationships were not precisely explained. One fundamental question is whether or not rill development is an essential sequel to the achievement of critical shear stress by runoff. LEOPOLD et al. (1966) found that 98% of the sediment produced in a semi-arid area of New Mexico over a ten year period was derived from unrilled hillslopes, while DUNNE (1980) emphasized the absence of rills from long hillslopes in Kenya, where SHIELDS sediment transport criteria were exceeded during rainstorms and particle movement was clearly apparent. In simulated rainfall experiments in Kenya (DUNNE & DIETRICH 1980a, b) and in Wyoming (EMMETT 1970) clear threads of flow concentration were apparent reflecting microtopographic irregularities and vegetation remnants, yet no rill incision was observed. In EMMETT's experiments in Wyoming, this was despite sprinkling intensities up to 266.7 mm/h for up to 6 hours! These results indicate that sheet erosion can occur without rill incision, and that flow concentration caused by microtopographic depressions does not necessarily lead to rill initiation and drainage system evolution, even if threshold conditions for particle entrainment are exceeded.

In addition to the field experiments described, carried out under simulated rainfall, two other studies are particularly relevant. SMITH & BRETHERTON (1972) used the continuity equation relating sediment transport and the rate of change of surface elevation to model the effect of introduced perturbations in the hillslope surface on the tendency of sheetwash to erode channels. This study has been fully discussed by DUNNE (1980). Essentially it showed that channel erosion will be expressed only if the initial slope profile is concave. The second study, by YAIR (1973), described rills along slope crests in Israeli badlands. This was interpreted as contradicting the existence of a belt without rilling, though it appears that such longitudinal rills will occur only if the crest is inclined, and are therefore consistent with HORTON's ideas. More interesting is the evidence that in this location rills can form at the apices of slopes which are convex in one or two dimensions, in contrast to SMITH & BRETHERTON's results.

EMMETT's (1970) study in Wyoming was one of the first attempts to study the hydraulics of overland flow in the field, and in comparison with precise laboratory experiments. It helped to stimulate a number of studies during the 1970's, which progressively characterized shallow flow hydraulics and the hydraulic conditions coincident with rill initiation. Particularly notable was the experimental work of SAVAT and DE PLOY in Belgium (SAVAT 1976, 1977, 1979, SAVAT & DE PLOEY 1982), KILING & RICHARDSON (1973) in the United States, and MOSS and his associates in Australia (MOSS & WALKER 1978, MOSS 1979, MOSS et al. 1979, 1980). SAVAT's studies confirmed and defined the occurrence of transitional laminar/turbulent flow on hillslopes, described by HORTON and EMMETT, demonstrated the existence of supercritical sheetflow on steep hillslopes, and examined the influence of rainsplash on sheetflow hydraulics. In experiments on highly erodible loess, SAVAT (1976) observed that rill initiation coincided with the developemnt of standing waves, which mould and deepen the bed channel, and later suggested that supercritical laminar flow is always associated with the development of roll waves or wave trains (SAVAT 1980). In further experiments, SAVAT & DE PLOEY (1982) found that rill initiation occurred with Fr Numbers of 2.8 (without rain) and 2.3 (with rain), though on very wet almost cohesionless material the threshold Fr Number dropped to 1.2. From these results they suggested that bed undulations would occur and deepen, once a threshold of Fr $1.2 + 0.003$ c' (c' = apparent cohesion expressed in $N/^m$), even if the original surface was plane and smooth. In their experiments this threshold was typically reached at slopes of 2–4°. This threshold was identified for Belgian calcareous loess, but in an extensive review of data from many field areas they found considerable agreement on a threshold angle of 2–3° for rill initiation on loamy soils, indicating the potentially broader significance of threshold Fr Numbers. Rather few data on Fr Numbers of thin flows are available, but KARCZ & KERSEY (1980) found that instability prevailed once Fr Numbers reached 0.5, while HODGES (1982) observed roll waves and incipient rill incision at Fr > 0.7 on gentle badland slopes, but also identified a higher threshold at Fr > 2.5 associated with the scouring of larger chutes. These were ephemeral features formed during rainstorms, which "healed" and disappeared with deposition during recession.

While SAVAT's most important contribution has been the characterization of sheetflow hydraulics, several other observations are also significant with regard to rill generation:

1. Rainsplash impact was found to have rather little effect on flow velocities and net drag force, but it always increased transport capacity and wash sediment concentrations, which in turn increased apparent kinematic viscosity by up to 30 per cent (SAVAT 1979). KILING & RICHARDSON (1973) also found that in extremely thin flows entrainment efficiency increased with rainfall rate. Rainsplash impact is therefore an important factor affecting particle entrainment in sheetflow and transport of particles from rill interfluve into channels.

2. Attempts to stimulate rill development with grooves or obstacles on the smooth loess surface were ineffective, and SAVAT concluded that rill initiation has little to do with vortex erosion.

3. Following studies of particle-size separation and the selectivity of rainsplash and sheetwash (SAVAT & POESEN 1977, SAVAT &

JANSSENS 1979, POESEN & SAVAT 1980, SAVAT 1982) it was concluded that rills form only if the hydraulic conditions allow coarse particles to be transported as easily as fine particles. In fact, laboratory experiments in small flumes (POESEN & SAVAT 1980) indicated that sheetwash and rill flow selectively transport coarse grains more effectively. MOSS et al. (1980) also noted the capacity of rills to stimulate pebble transport down gentle slopes, and POESEN (this volume) has demonstrated gravel transport by rill flow at the Huldenberg Experimental Site.

While the research described above was in progress, MOSS and his associates were involved in similar work in Australia (MOSS & WALKER 1978, MOSS et al. 1979, 1982). This originated in sedimentological study of bedload transport in rivers (MOSS 1972) and MOSS & WALKER (1978) demonstrated the widespread occurrence of analogous deposits as evidence of bedload transport on hillslopes. Clear distinction between suspended transport of fines and bedload transport of coarse particles was possible in many of the overland flows observed, though rheologic transport of mixed material was also noted. Most flows were supercritical and turbulent, and it was observed that channels were only incised when the hydraulic threshold conditions for bedload transport occurred, confirming the requirement of nonselective erosion observed in the Belgian studies. Such channels would generally be regarded as rills, but MOSS & WALKER (1978) adopted an unusual definition, using the term "rill" only for channels in which active deposition is taking place, as a result of decreased capacity. They developed the concept of hillslope erosion controlled by a "hydraulic mantle" of bedload material (more usually referred to as "armouring"), and the interesting concept of "stored potential for hydraulic adjustment" as evidenced by, for example, well-developed soil profiles.

This broad conceptual study was followed by detailed experimental studies of overland flow sediment transport processes. MOSS et al. (1979) confirmed SAVAT's (1977) observations on the interaction between flow depth and the effectiveness of rainsplash impact, and showed the effect of splash-induced "rainflow" transport (described by DE PLOEY (1971) as "rainwash") on channel initiation. Where this is active, flow channelization and rill initiation are suppressed and sheetflow predominates, in association with transverse ripples. The effectiveness of rainwash depends closely on flow depth, and diminishes downslope as flow depth increases, until suitable conditions for channel initiation are established. The effect of drop-size is less well-known, and MOSS et al.'s (1979) description of the way in which segregation of large drops into short "bursts", and the associated rapid fluctuations in rainwash intensity, can contribute to the development of roll-waves and channel initiation is interesting.

In other studies (MOSLEY 1973, MOSS et al. 1980, 1982) the location of rill incision has been related to flowlines in sheetflow. In MOSS et al.'s (1982) view, these "protochannels" were associated with the development of dominant secondary flows, directed towards the bed and causing lateral transport of sediment and the evolution of micro-levees. In discharge conditions close to the threshold for bedload transport, these protochannels would form spontaneously on smooth beds, but they could also be stimulated at much lower discharges by introducing objects into the flow, provided the flow width exceeded the depth by a factor of about ten. Small grooves in the bed were also effective in stimulating protochannel development, contrary to the findings of SAVAT, but the reasons for this discrepancy are not clear.

MOSS et al. (1982) suggested that the effect of raindrop impact was to suppress the development of secondary flow cells and flow instability, and therefore to suppress channel initiation. This conflicts with SAVAT's (1979) observation that the increased sediment loads and apparent kinematic viscosity, associated with rainsplash, enhanced the potential for rill initiation. However, both studies showed a fundamental change in the processes active once the flow is sufficiently deep to protect the soil surface from raindrop impact. These processes are usually distinguished as "rill" and "interrill" processes, using the terminology introduced by MEYER et al. (1975) in order to identify the different locations of sediment sources within the soil profile. Sediment transported by interrill erosion is typically entrained from the topsoil by rainsplash, while sediment entrainment within rills is derived from subsurface horizons or parent material. YOUNG & ONSTAD (1978) developed laboratory instrumentation to separate the rill and interrill erodibility of soils, and several studies have shown that interrill flow is less effective in transporting large particles than rill flow (ALBERTS et al. 1980, LOCH & DONNELLAN 1983a, b).

While rill erosion has received comparatively little attention, interrill erosion has been intensively studied since the pioneer work of YODER (1936), DULEY (1939) and ELLISON (1947), amongst others. Studies have focussed on the detailed interaction of rainsplash and sheetwash entrainment processes (e.g. DE PLOEY 1971, YOUNG & WIERSMA 1973, BOLLINNE 1978, MORGAN 1978, DE PLOEY & MÜCHER 1981, POESEN 1981, SAVAT & POESEN 1981, POESEN & SAVAT 1981), measurement of rainwash erosion rates under simulated rainfall in the laboratory or field in relation to slope angle (e.g. BRYAN 1969, 1974, 1979, MOEYERSONS & DE PLOEY 1976, DE MEESTER et al. 1979, SINGER et al. 1982), determination of the soil properties influencing erodibility (ADAMS et al. 1958, BRYAN 1968, BRUCE-OKINE & LAL 1975, DE VLEESCHAUWER et al. 1978, BARBER et al. 1979, SINGER et al. 1980, BRYAN & DE PLOEY 1983) and identification of soil crusting processes (e.g. HILLEL 1960, EPSTEIN & GRANT 1967, FARRES 1978, MORIN et al. 1981, DE PLOEY 1981). Comprehensive reviews have been provided by SMITH & WISCHMEIER (1962) and DE PLOEY & POESEN (1984). These show that the processes and rates of interrill erosion are extremely complex and varied, depending both on extrinsic factors, such as rainfall intensity, raindrop size and the presence or absence of wind influence, and intrinsic factors, such as soil texture, aggregation characteristics, surface roughness, susceptibility to crusting, and the presence and density of organic debris.

Recognition of the importance of distinguishing rill and interrill erosion processes has lead to attempts to incorporate these in physically-based models of hillslope erosion processes. FOSTER (1982) has provided an extensive review of the evolution of such models, particularly in the United States, while ROSE (1985) has reviewed more recent material, with particular emphasis on research carried out in Australia. Some of this work is also reviewed by LOCH & THOMAS (this volume) as a background to their study of rill development processes on swelling clay soils in Queensland. In another recent study of rill evolution on hillslope plots under simulated rainfall in southern Kenya, DUNNE & AUBRY (1986) found that the dynamics of the rill system could be explained by the balance in intensity of rill and interrill processes, or sheetwash and rainsplash processes.

In the absence of rainsplash, sheetflow became unstable and incised rills, but when rainsplash was added, splash transport of sediment from interrills tended to fill and eliminate rills. The precise morphologic expression of the rill system will vary in space and time, reflecting fluctuations in the relative effectiveness of these processes. Variations in space reflect particularly the increased effectiveness of sheetwash with slope angle, while rainsplash is particularly sensitive to the depth of the surface water layer, and will tend to decrease systematically downslope as this depth increases. The precise balance of these effects will vary greatly with rainstorm characteristics, and this will tend to be reflected in the morphologic expression of the rill system.

The conceptual framework developed by DUNNE & AUBRY (1986) is important because it emphasizes the dependance of the rill system on interrill processes as well as on the hydraulic characteristics of flow in rill channels. Interrill processes are strongly controlled by the properties of the surface soils which are usually much more complex and varied than those of the subsoil which is directly involved in the later stages of rill development. DUNNE & AUBRY's (1986) field experiments were carried out on relatively coarse-textured soils with poor aggregation, relatively low cohesion and limited horizon differentiation. Most of the soils studied by SAVAT and by MOSS had similar characteristics, allowed the rill system to respond swiftly and sensitively to changes in storm conditions, and show the rill-interrill model operating in what is probably its simplest form. Even these conditions frequently became more complex as a result of the progressive development of armouring on their experimental plots, and ROWNTREE (1982) has also shown this in laboratory experiments under simulated rainfall. The precise development and effects of armouring vary, but typically it appears first on interrill areas due to selective transport. This progressively reduces splash diffusion from interrills, and usually reduces interrill infiltration capacities as well, so that less sediment and more water enter the rill channel, leading to enhanced incision. Often the rill channel will incise through to subsurface horizons which, with lower organic contents and often poorer aggregation, are frequently more erodible, so that incision of the rill system is accompanied by rapid increase in the overall rate of soil loss. In plot experiments under simulated rainfall, MEYER et al. (1975) found a threefold increase in the rate of soil loss, coincident with rill development. Apart from any geomorphic significance this demonstrates the clear agricultural importance of rill development.

Most experimental evidence on the integrated application of the rill-interrill model relates to conditions in which it operates in a relatively simple form. There is a clear need to expand experimental testing to determine if this conceptually simple model can be applied to a wider range of soil conditions. Experiments on fine-grained cohesive soils with varied chemical characteristics are particularly important, so that the full complex range of aggregation dynamics can be incorporated in general concepts of rill evolution. The experiments by LOCH & THOMAS on cracking clays in Queensland (this volume) are significant contributions in this direction. The simple model must become much more complicated, even when applied to rill evolution during a single storm, as it must be capable of accomodating time-dependent changes in rainsplash transport related to differential aggregate disintegration and surface crust evolution. In order to apply also to long-term rill evolution on an individual hillslope, the model must also be capable of accomodating changes in the character of aggregation and surface crusts related to the ageing characteristics of clays (GRISSINGER 1966), antecedent soil moisture conditions (COUSEN & FARRES 1984), and progressive changes in soil organic matter properties (TISDALL & OADES 1982). Some changes are physical and many research projects have examined physical changes in aggregation and surface crust behaviour and their relationship to soil constituents. However, an increasing number of studies (e.g. SARGUNAM et al. 1973, ARULANANDAN et al. 1975, IMESON et al. 1982, IMESON & VERSTRATEN 1985, GERITS et al. 1986) have made it clear that important changes in both aggregation and crust characteristics can result from changes in the balance of soil and pore fluid chemistry. Some changes are comparatively long-term in effect, but chemical conditions can also produce prefound changes in the erodibility of surface soils within a single rainstorm (GERITS 1986).

Any comprheensive model of rill evolution must incorporate the effect of soil and pore fluid chemistry as well as purely physical influences. Most of the changes in aggregation and crusting referred to above involve rather complex physico-chemical processes, but purely chemical processes of solution may also dominate rill development. In this volume FORD & LUNDBERG provide a comprehensive review of channelized solution features developed on a variety of bed rock types, together with information from recent experimental studies. Traditionally, study of solution features such as "rillenkarren" has been related to karst geomorphology and calcareous rocks and has not been considered as part of the more general phenomenon of rill development, but recent studies of rill evolution on smectitic siltstones in the Dinosaur Badlands of western Canada (BOWYER-BOWER & BRYAN 1986) suggest that the initiation of rills on some non-calcareous rocks can also be dominated by solution.

The rill-interrill model treats rill evolution as being controlled by surface erosion and deposition processes. This is probably appropriate for the soils which have been studied in connection with rill development, but expansion of research to smectite-rich rocks has revealed a completely different pattern of rill evolution. These rocks have very high shrink-swell capacities and surfaces are dominated by desiccation cracking. At the start of rainfall, crack infiltration capacities are very high, preventing surface runoff. Flow starts as subsurface and crack flow, which emerges to the surface only at depressions where silt deposition may occur. Rill evolution proceeds in two ways. Patches of silt deposition become preferential sites for initiation of surface runoff and rill initiation (BRYAN et al. 1978), while in highly dispersive conditions zones of crack flow and subsurface flow evolve into micropipes and tunnels, which eventually expand the rill system by roof collapse (HODGES & BRYAN 1982). The initial stages of rill evolution on these materials are not related to flow hydraulics but are controlled by moisture concentration and its effects on material properties, and by the channel conditions conducive to dispersion. Origin from subsurface flow would appear to explain the rill development on convex interfluves in Israel, described by YAIR (1973), which appeared to conflict with HORTON's concepts and SMITH & BRETHERTONS's (1972) mathematical model. In this volume GERITS et al. have described the evolution of rills from micropipes and tunnels in Spain and Canada, and have shown that the relationship between rill spacing and soil properties may be described by the same expression used for the spacing of field tile drains. It is clear that a pipe model of rill development is much more applicable to swelling clays than the orthodox rill-interrill model, but it probably has much wider application to cohesive, well-aggregated soils which retain high infiltration capacities except under extreme rainfall conditions. The model seems particularly applicable to agricultural soils of good tilth with rough surfaces created by tillage. On non-agricultural soils the evolution of pipe-systems may be more strongly influenced by root development and animal burrows (JONES 1981), and in this volume GOVERS describes mole burrows as an important influence on the development of rills at the Huldenberg Experimental Site.

On hillslopes in the Dinosaur Badlands of western Canada at least three different processes are involved in rill initiation. On smectitic mudstones rills originate in micropiping and differential silt deposition, as described above, and sometimes also in slope instability and mudflows (BRYAN et al. 1978). On micropediments rills develop primarily in response to the hydraulic condition of surface flow (HODGES 1982), while on resistant siltstones the dominant process appears to be solution. On most hillslopes thin siltstones and micropediments are interbedded with mudstones, so that the evolution of the complete hillslope rill system and the processes of runoff initiation are complex and discontinuous. Badland hillslopes are extreme examples of the controlling influence of lithology on rill evolution, but differences in soil properties and bedrock characteristics are sufficiently marked to ensure that discontinuous rill evolution is widespread and should be recognized in any general consideration of the contribution of rills to drainage system evolution. In this volume PLANCHON et al. provide an excellent example of discontinuous rill development in hillslopes in the wet savannah region of Côte d'Ivoire. These hillslopes occur over homogeneous gneiss, but marked differences in soil properties have developed due to prolonged pedogenesis and discontinuous rill evolution is closely related to soil catena characteristics. Three distinct rill systems occur which are linked by integrated flow only at the end of the rainy season, when the rising water table obscures pedological differences. In the completely different and much more arid environment of the Negev Desert, Israel, WEIDER et al. (1985) and YAIR & LAVEE (1985) have also demonstrated the relationship of soil catenas to discontinuous drainage development.

Any significant lithological or pedological differentiation will strongly influence rill evolution, but experimental work in Belgium has also shown that discontinuities in rill evolution can also arise due to channel hydraulic conditions. Although his earlier work related rill initiation to the flow Froude Numbers, SAVAT (1982) found that the mean particle size transported from loess soils was closely related to flow shear velocity, and transport reached the aselective condition which he associated with rill development at a critical threshold shear velocity of 3.0 cm/s. In tests with Tongrian sand and a variety of silt loam and sandy loam soils, GOVERS (1985) found that critical shear velocities typically ranged from 3.0 to 3.5 cm/s and that on loamy soils these figures were associated with slope angles of 2–3°, agreeing with SAVAT & DE PLOEY's (1982) data, derived from literature review. When the slope angle falls below the threshold for the critical shear velocity, transport will again become selective and colluvial deposition will start in the rill channel. DE PLOEY (1984) also developed an empirical expression for determining the critical angle at which colluviation starts (S_{cr}) as:

$$Sr = A, c^{08.}/q^{0.5} \qquad (1)$$

where
c   = flow load concentration,
q   = flow unit discharge in cm²/s,
S_{cr} = tan S_{cr},
A   = an empirical coefficient, a function of the mean grain diameters.

GOVERS & RAUWS (1986) found that this expression underestimated the transport capacity of thin flows, possibly because it describes the onset of deposition of the coarsest particles, while the transport capacity for smaller particles is still not fully satisfied.

Some of the recent studies discussed briefly above have shown that rill initiation in certain situations can be described by one of the developments of the comparatively simple hydraulic model originally conceived by HORTON (1945). There is still much to be learned about the hydraulics of thin layer flows and the details of their influence on rill development. HORTON indicated the potential significance of instability features such as roll waves in rill development and RAUWS' paper, in this volume, reports one of the first attempts to examine this by formal experimental studies on non-cohesive soils. Soil cohesion adds another complex dimension and the paper by TORRI et al., in this volume, reports initial experiments on its significance. The rill-interrill model can be greatly refined by continued study along these lines, and by careful integration of the results of many studies of interrill processes. However, it is clear from some of the recent studies discussed that the full range of rill development processes in very diverse soil conditions cannot be adequately encompassed by a physical surface erosion model alone. In fact, it appears that the importance of hydraulic action has been greatly overemphasized and that the conditions appropriate to the surface "drainage" model of GERITS et al. (this volume) are much more widespread, particularly on well-developed or agricultural soils. Any comprehensive concept of rill development must include rill initiation by subsurface flow and the influence of chemical and physico-chemical processes. It is increasingly clear that it must also include biological processes and the time-dependent influence of organic decomposition on soil properties. Significant progress in understanding the practical and geomorphic significance of rill development can only occur if the controlling influence of soil properties is fully appreciated. An extensive series of field and laboratory experiments on a wide range of soil types will be required to determine the limiting pedologic conditions for the application of different models of rill initiation. Common agreement on a more precise definition of the term "rill" would be a useful adjunct of such experimental research.

REFERENCES

[ADAMS et al. 1958] ADAMS, J., KIRKHAM, D. & SHOLTES, W.: Soil erodibility and other physical properties of some Iowa soils. Iowa State College Journal of Science, **32**, 485–540.

[ALBERTS et al. 1980] ALBERTS, E., MOLDENHAUER, W. & FOSTER, G.: Soil aggregates and primary particles transported in rill and interrill flow. Soil Science Society of America Journal, **44**, 590–595.

[ARULANANDAN et al. 1975] ARULANANDAN, K., LOGANATHAN, P. & KRONE, R.B.: Pore and eroding fluid influences on surface erosion of soil. Journal of the Geotechnical Engineering Division, American Society of Civil Engineering, **101**, GT 1, 51–66.

[BARBER et al. 1979] BARBER, R.G., MOORE, T.R. & THOMAS, D.: The erodibility of two soils from Kenya. Journal of Soil Science, **30**, 579–591.

[BOLLINNE 1978] BOLLINNE, A.: Study of the importance of splash and wash on cultivated loamy soils of Hesbaye(Belgium). Earth Surface Processes, **3**, 71–84.

[BOWYER-BOWER & BRYAN 1986] BOWYER-BOWER, T.A.S. & BRYAN, R.B.: Rill initiation: concepts and evaluation on badland slopes. Zeitschrift für Geomorphologie, Supplement Band, **59**, 161–175.

[BRUCE-OKINE & LAL 1975] BRUCE-OKINE, E. & LAL, R.: Soil erodibility as determined by raindrop technique. Soil Science, **119**, 149–157.

[BRYAN 1968] BRYAN, R.B.: Development, use and efficiency of indices of soil erodibility. Geoderma, **2**, 5–26.

[BRYAN 1969] BRYAN, R.B.: The relative erodibility of soils developed in the Peak District of Derbyshire. Geografiska Annaler, **51A**, 145–159.

[BRYAN 1974] BRYAN, R.B.: Water erosion by splash and wash and the erodibility of Alberta soils. Geografiska Annaler, **56A**, 159–181.

[BRYAN 1979] BRYAN, R.B.: The influence of slope angle on soil entrainment by sheetwash and rainsplash. Earth Surface Processes, **4**, 43–58.

[BRYAN et al. 1978] BRYAN, R.B., YAIR, A. & HODGES, W.K.: Factors controlling the initiation of runoff and piping in Dinosaur Provincial Park badlands, Alberta, Canada. Zeitschrift für Geomorphologie, Supplement Band **29**, 151–168.

[BRYAN & PRICE 1980] BRYAN, R.B. & PRICE, A.G.: Recession of the Scarborough Bluffs, Ontario, Canada. Zeitschrift für Geomorphologie, Supplement Band **34**, 48–62.

[BRYAN & DE PLOEY 1983] BRYAN, R.B. & DE PLOEY, J.: Comparability of soil erosion measurements with different laboratory rainfall simulators. In: Rainfall Simulation, Runoff and Soil Erosion. J. DE PLOEY (Ed), CATENA Supplement, **4**, 33–56.

[COUSEN & FARRES 1984] COUSEN, S.M. & FARRES, P.J.: The role of moisture control in the stability of soil aggregates from a temperate silty soil to raindrop impact. CATENA **11**, 313–330.

[DE MEESTER et al. 1979] DE MEESTER, T., IMESON, A.C. & JUNGERIUS, P.D.: Some problems in assessing soil loss from small-scale field measurements. In: Soil Physical Properties and Crop Production in the Tropics. R. LAL & D. GREENLAND (Eds.), 466–473, Wiley, Chichester.

[DE PLOEY 1971] DE PLOEY, J.: Liquefaction and rainwash erosion. Zeitschrift für Geomorphologie, **15**, 491–496.

[DE PLOEY 1981] DE PLOEY, J.: Crusting and time-dependent rainwash mechanisms on loamy soil. In: Soil Conservation. R.P.C. MORGAN (Ed.), 139–154, Wiley, Chichester.

[DE PLOEY 1984] DE PLOEY, J.: Hydraulics of runoff and loess loam deposition. Earth Surface Processes and Landforms, 9, 523–531.

[DE PLOEY & MÜCHER 1981] DE PLOEY, J. & MÜCHER, H.J.: A consistency index and rainwash mechanisms on Belgian loamy soils. Earth Surface Processes and Landforms, 6, 319–330.

[DE PLOEY & POESEN 1984] DE PLOEY, J. & POESEN, J.: Aggregate stability, runoff generation and interrill erosion. In: Geomorphology and Soils. K. RICHARDS & S. ELLIS (Eds.), George Allen & Unwin, London.

[DE VLEESCHAUWER et al. 1978] DE VLEESCHAUWER, D., LAL, R. & DE BOODT, M.: Comparison of detachability indices in relation to soil erodibility for some important Nigerian soils. Pédologie, 28, 5–20.

[DULEY 1939] DULEY, F.L.: Surface factors affecting the rate of water intake by soils. Soil Science Society of America Proceedings, 4, 60–64.

[DUNNE 1980] DUNNE, T.: Formation and control of channel networks. Progress in Physical Geography, 4, 211–239.

[DUNNE & DIETRICH 1980a] DUNNE, T. & DIETRICH, W.E.: Experimental study of Horton overland flow on tropical hillslopes. 1. Soil conditions, infiltration and frequency of runoff. Zeitschrift für Geomorphologie, Supplement Band, 35, 40–59.

[DUNNE & DIETRICH 1980b] DUNNE, T. & DIETRICH, W.E.: Experimental study of Horton overland flow on tropical hillslopes. 2. Hydraulic characteristics and hillslope hydrographs. Zeitschrift für Geomorphologie, Supplement Band, 35, 60–80.

[DUNNE & AUBRY 1986] DUNNE, T. & AUBRY, B.F.: Evaluation of Horton's theory of sheetwash and rill erosion on the basis of field experiments. In: Hillslope Processes. A.D. ABAHAMS (Ed.), Allen & Unwin, Boston, 31–53.

[ELLISON 1947] ELLISON, W.D.: Soil erosion studies. Journal of Agricultural Engineering, 28, 145–146, 197–201, 297–300, 349–351.

[EMMETT 1970] EMMETT, W.W.: The hydraulics of overland flow on hillslopes. United States Geological Survey Professional Paper, 662-A.

[EPSTEIN & GRANT 1967] EPSTEIN, E. & GRANT, W.: Soil losses and crust formation as related to some soil physical properties. Soil Science Society of America Proceedings, 31, 547–550.

[F.A.O. 1965] F.A.O.: Soil Erosion by Water: Some Measures for its Control on Cultivated Lands. FAO/UNESCO, Rome.

[FARRES 1978] FARRES, P.: The role of time and aggregate size in the crusting process. Earth Surface Processes, 3, 243–254.

[FOSTER 1982] FOSTER, G.: Modelling of the erosion process. In: Hydrologic Modelling of Small Watersheds. C.T. HAAN (Ed.): American Society of Agricultural Engineers Monograph, 5.

[FOSTER & MEYER 1975] FOSTER, G. & MEYER, L.: Mathematical simulation of upland erosion by fundamental erosion mechanics. In: Present and Prospective Technology for Predicting Sediment Yields and Sources. Agricultural Research Service Report hbARS-S-40, 190–207.

[GERITS 1986] GERITS, J.J.P.: Implications of chemical thresholds and physico-chemical processes for modelling erosion in southeastern Spain. Paper to Commission on Measurement, Theory and Application in Geomorphology, Granada.

[GERITS et al. 1986] GERITS, J.J.P., IMESON, A.C. & VERSTRATEN, J.M.: Chemical thresholds and erosion in saline and sodic materials. In: Estudios sobre Geomorfologia del Sur de Espana. F. LOPEZ BERMUDEZ & J.B. THORNES (Eds.), 75–79. University of Murcia, Murcia.

[GOVERS 1985] GOVERS, G.: Selectivity and transport capacity of thin flows in relation to rill erosion. CATENA 12, 35–50.

[GOVERS & RAUWS 1986] GOVERS, G. & RAUWS, G.: Transporting capacity of overland flow on plane and irregular beds. Earth Surface Processes and Landforms, 11, 515–524.

[GRISSINGER 1966] GRISSINGER, E.H.: Resistance of selected clay systems to erosion by water. Water Resources Research, 2, 131–138.

[HILLEL 1960] HILLEL, D.: Crust formation in loessial soils. Proceedings of the Seventh International Congress of Soil Science, 330–339.

[HORTON 1945] HORTON, R.E.: Erosional development of streams and their drainage basins: hydrophysical approach to quantitative morphology. Bulletin of the Geological Society of America, 56, 275–370.

[HODGES 1982] HODGES, W.K.: Hydrologic characteristics of a badland pseudopediment slope system during simulated rainstorm experiments. In: Badland Geomorphology and Piping. R.B. BRYAN & A. YAIR (Eds.), 127–152. GeoBooks, Norwich.

[HODGES & BRYAN 1982] HODGES, W.K. & BRYAN, R.B.: The influence of material behaviour on runoff initiation in the Dinosaur Badlands, Canada. In: Badland Geomorphology and Piping. R.B. BRYAN & A. YAIR (Eds.), 13–46, GeoBooks, Norwich.

[IMESON et al. 1982] IMESON, A.C., KWAAD, F.J.P.M. & VERSTRATEN, J.M.: The relationship of soil physical and chemical properties to the development of badlands in Morocco. In: Badland Geomorphology and Piping. R.B. BRYAN & A. YAIR (Eds.), 47–70. GeoBooks, Norwich.

[IMESON & VERSTRATEN 1985] IMESON, A.C. & VERSTRATEN, J.M.: The erodibility of highly calcareous soil material from southern Spain. CATENA **12**, 291–306.

[JONES 1981] JONES, J.A.A.: The Nature of Soil Piping—A Review of Research. GeoBooks, Norwich.

[KARCZ & KERSEY 1980] KARCZ, I. & KERSEY, D.: Experimental study of freesurface flow instability and bedforms in shallow flows. Sedimentary Geology, **27**, 263–300.

[KILINC & RICHARDSON 1973] KILINC, M. & RICHARDSON, E.V.: Mechanics of soil erosion from overland flow generated by simulated rainfall. Colorado State University Hydrology Paper, **63**.

[LEOPOLD et al. 1966] LEOPOLD, L.B., EMMETT, W.W. & MYRICK, R.M.: Channel and hillslope processes in a semi-arid area. United States Geological Survey Professional Paper, **252**.

[LOCH & DONNELLAN 1983a] LOCH, R.J. & DONNELLAN, T.E.: Field simulator studies on two clay soils of Darling Downs, Queensland. I. The effect of plot length and tillage orientation on erosion processes and runoff and erosion rates. Australian Journal of Soil Research, **21**, 33–46.

[LOCH & DONNELLAN 1983b] LOCH, R.J. & DONNELLAN, T.E.: Field simulator studies on two clay soils of Darling Downs, Queensland. II. Aggregate breakdown, sediment properties and soil erodibility. Australian Journal of Soil Research, **21**, 47–58.

[MEYER et al. 1975] MEYER, L., FOSTER, G. & RÖMKENS, M.: Source of soil eroded by water from upland slopes. In: Present and Prospective Technology of Predicting Sediment Yields and Sources. Agricultural Research Service Report, **ARS-S-40**, 177–189.

[MOEYERSONS & DE PLOEY 1976] MOEYERSONS, J. & DE PLOEY, J.: Quantitative data on splash erosion, simulated on unvegetated slopes. Zeitschrift für Geomorphologie, Supplement Band, **25**, 120–131.

[MORGAN 1978] MORGAN, R.P.C.: Field studies of rainsplash erosion. Earth Surface Processes, **3**, 295–299.

[MORIN et al. 1981] MORIN, J., BENYAMINI, Y. & MICHAELI, A.: The effect of raindrop impact on the dynamics of soil surface crusting and water movement in the profile. Journal of Hydrology, **52**, 321–335.

[MOSLEY 1973] MOSLEY, M.P.: An experimental study of rill erosion. Transactions of the American Society of Agricultural Engineers, **17**, 909–913.

[MOSS 1972] MOSS, A.J.: Bedload sediments. Sedimentology, **18**, 159–219.

[MOSS & WALKER 1978] MOSS, A.J. & WALKER, P.H.: Particle transport by continental water flows in relation to erosion, deposition, soils and human activities. Sedimentary Geology, **20**, 81–139.

[MOSS 1979] MOSS, A.J.: Thin flow transport of solids in arid and non-arid areas: a comparison. International Association of Hydrological Sciences, Symposium on Hydrology of Areas of Low Precipitation, Canberra. International Association of Hydrological Sciences Publication, **128**, 435–445.

[MOSS et al. 1980] MOSS, A.J., WALKER, P.H. & HUTKA, J.: Movement of loose sandy detritus by shallow water flows: an experimental study. Sedimentary Geology, **25**, 43–66.

[MOSS et al. 1982] MOSS, A.J., GREEN, P. & HUTKA, J.: Small channels: their experimental formation, nature and significance. Earth Surface Processes and Landforms, **7**, 401–416.

[POESEN 1981] POESEN, J.: Rainwash experiments on the erodibility of loose sediments. Earth Surface Processes and Landforms, **6**, 285–302.

[POESEN & SAVAT 1980] POESEN, J. & SAVAT, J.: Particle size separation during erosion by splash and runoff. In: Assessment of Erosion. M. DE BOODT & D. GABRIELS (Eds.), 427–439. Wiley, chichester.

[POESEN & SAVAT 1981] POESEN, J. & SAVAT, J.: Detchment and transportation of loess sediments by raindrop splash. Part II. Detachability and transportability measurements. CATENA, **8**, 19–41.

[ROSE 1985] ROSE, C.W.: Developments in soil erosion and deposition models. In: Advances in Soil Science, **2**. B.A. STEWART (Ed.), Springer Verlag, Berlin.

[ROWNTREE 1982] ROWNTREE, K.M.: Sediment yields from a laboratory catchment and their relationship to rilling and surface armouring. Earth Surface Processes and Landforms, **7**, 153–170.

[SARGUNAM et al. 1973] SARGUNAM, A., RILEY, P., ARULANANDAN, K. & KRONE, R.B.: Physico-chemical factors in erosion of cohesive soils. Journal of the Hydraulics Division, American Society of Civil Engineers, **99**, HY3, 555–558.

[SAVAT 1976] SAVAT, J.: Discharge velocities and total erosion of a calcareous loess: a comparison between pluvial and terminal runoff. Revue Géomorphologie Dynamique, **24**, 113–122.

[SAVAT 1977] SAVAT, J.: The hydraulics of sheet flow on a smooth surface and the effect of simulated rainfall. Earth Surface Processes, **2**, 125–140.

[SAVAT & POESEN 1977] SAVAT, J. & POESEN, J.: Splash and discontinuous runoff as creators of fine sandy lag deposits with Kalahari sands. CATENA, **4**, 321–332.

[SAVAT 1979] SAVAT, J.: Laboratory experiments on erosion and deposition of loess by laminar sheetflow and turbulent rillflow. In: Proceedings of Seminar Agricultural Soil Erosion in Temperate Non-Mediterranean Climates. H. VOGT & T. VOGT (Eds.), Institut Louis Pasteur, Strassbourg-Colmar.

[SAVAT & JANSSEN 1979] SAVAT, J. & JANSSEN, V.: Experimenten betreffende de selectiviteit van runoff en spat. Bulletin Société Géologique de Belgique, **88**, I–II, 113–122.

[SAVAT 1980] SAVAT, J.: Resistance to flow in rough supercritical sheetflow. Earth Surface Processes, **5**, 103–122.

[SAVAT & POESEN 1981] SAVAT, J. & POESEN, J.: Detachment and transportation of loose sediments by raindrop splash. Part I. The calculation of absolute data on detachability and transportability. CATENA, **8**, 1–17.

[SAVAT 1982] SAVAT, J.: Common and uncommon selectivity in the process of fluid transportation: field observations and laboratory experiments on bare surfaces. In: Aridic Soils and Geomorphic Processes. D.H. YAALON (Ed.), CATENA SUPPLEMENT, **1**, 139–160.

[SAVAT & DE PLOEY 1982] SAVAT, J. & DE PLOEY, J.: Sheetwash and rill development by surface flow. In: Badland geomorphology and Piping. R.B. BRYAN & A. YAIR (eds.), 113–126. GeoBooks, Norwich.

[SCHUMM 1956] SCHUMM, S.A.: Evolution of drainage systems and slopes in badlands at Perth Amboy, New Jersey. Bulletin of the Geological Society of America, **67**, 597–646.

[SCHUMM & LUSBY 1963] SCHUMM, S.A. & LUSBY, G.C.: Seasonal variation of infiltration capacity and runoff on hillslopes in western Colorado. Journal of Geophysical Reaearch, **68**, 3655–3666.

[SINGER et al. 1980] SINGER, M., BLACKARD, J. & JANITSKY, P.: Dithionite iron and soil cation content as factors in soil erodibility. In: Assessment of Soil Erosion. M. DE BOODT & D. GABRIELS (Eds.), 259–267, Wiley, Chichester.

[SMITH & WISCHMEIER 1962] SMITH, D.D. & WISCHMEIER, W.H.: Rainfall erosion. Advances in Agronomy, **14**, 109–148.

[SMITH & BRETHERTON 1972] SMITH, T.R. & BRETHERTON, F.P.: Stability and the conservation of mass in drainage basin evolution. Water Resources Research, **8**, 1506–1529.

[TISDALL & OADES 1982] TISDALL, J.M. & OADES, J.M.: Organic matter and water-stable aggregates in soils. Journal of Soil Science, **33**, 141–163.

[WEIDER et al. 1985] WEIDER, M., YAIR, A. & ARZI, A.: Catenary soil relationships on arid hillsopes. In: Soils and Geomorphology. P.D. JUNGERIUS (Ed), CATENA SUPPLEMENT, **6**, 41–57.

[YAIR 1973] YAIR, A.: Theoretical considerations on the evolution of convex hillslopes. Zeitschrift für Geomorphologie, Supplement Band, **18**, 1–9.

[YAIR & LAVEE 1985] YAIR, A. & LAVEE, H.: Runoff generation in arid and semi-arid zones. In: Hydrological Forecasting. M.G. ANDERSON & T.P. BURT (Eds.), 183–220, Wiley, Chichester.

[YODER 1936] YODER, R.E.: A direct method of aggregate analysis of soils and a study of the physical nature of erosion losses. Journal of the American Society of Agronomy, **28**, 337–351.

[YOUNG & WIERSMA 1973] YOUNG, R. & WIERSMA, J.: The role of rainfall impact in soil detachment and transport. Water Resources Research, **9**, 1629–1636.

[YOUNG & ONSTAD 1978] YOUNG, R. & ONSTAD, C.A.: Characteristics of rill and interrill eroded soils. Transactions of the American Society of Agricultural Engineers, **21**, 1126–1130.

Address of author:
R.B. Bryan
Department of Geography, University of Toronto (Scarborough Campus),
1265, Military Trail,
Scarborough, Ontario L3P 3B5,
Canada

SPATIAL AND TEMPORAL VARIABILITY IN RILL DEVELOPMENT PROCESSES AT THE HULDENBERG EXPERIMENTAL SITE

G. **Govers**, Leuven

SUMMARY

In this paper the evolution of a rill and gully system during a one year period on an experimental field occupying a whole slope section is described and analysed. The experimental field of Huldenberg is characterised by a pronounced spatial variability of lithological and pedological properties allowing observation of a number of phenomena.

Basic information was acquired through cartographic surveys of the rill and gully system. Observations made it clear that it was necessary to distinguish between hydraulic rill erosion, mass wasting processes on rill sidewalls and gullying. Attention was also paid to the occurrence of soil piping.

The main factors controlling the spatially varied evolution of these processes are discussed. Hydraulic rill erosion is strongly controlled topographically and occurs mainly during major runoff events. Mass movement processes are much more dependent on the mechanical and hydrological properties of the soil profile. Gully erosion only occurs where the subsoil consists of loose sand. Piping is restricted to zones where a topsoil with reasonable structural characteristics overlies a compact impervious subsoil and is often initiated by biological activity. On arable land, many erosion processes are thus controlled not only by the characteristics of the topsoil, but by the properties of the whole soil profile.

1 INTRODUCTION

During the past two decades, research on soil erosion on arable land has increasingly focussed on the separate study of the different processes involved. A number of so-called physically-based erosion models and a limited amount of experimental work on rill and interrill erosion have resulted. However, most field observations are still carried out on classical plots, typically occupying a slope segment of limited length (e.g. BOLLINNE 1982). In most cases, only total runoff and sediment output are measured, while the relative contribution of the individual processes is not considered. As the processes are not considered separately, no information is gained about the factors controlling them or their interelationships. Furthermore, the development

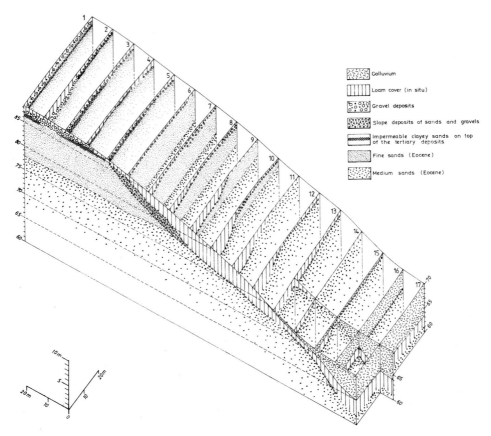

Figure 1: *Surficial deposits of the Huldenberg field (after E. PAULISSEN).*

of specifically rill erosion on plots of only limited length will always be constrained so that some phenomena which play an important role in rill development on a normal slope may not be observed.

These deficiencies can be overcome by studying rill development on a whole slope section, whereby attention is focussed on the spatially and temporally differentiated evolution within the field and not on total sediment output. This paper describes the results of such a study of rill development on the Huldenberg experimental field during one year. An attempt has been made to clarify the evolution of the whole system by considering the different processes contributing to rill and gully erosion. Adequate information on rill development was acquired by volumetric recording of the rill pattern development, a technique which has been little used (BOARDMAN 1984, BOLLINE 1977, FULLEN 1985, GABRIELS et al. 1977). To our knowledge, only MC COOL & GEORGE (1983) used this technique to obtain systematic information on rill development as influenced by different factors.

2 MATERIALS AND METHODS

The experimental field is located in Huldenberg, in the loamy hilly region of Flanders, 15 km southwest of Leuven. It occupies a 0.75 ha south-facing convex-concave slope section (fig.1). The maximum slope amounts to 14 degrees, which is rather high compared with other cultivated fields in the region. The characteristics of the surficial deposits, which were investigated by PAULISSEN (pers. comm.) are very variable. A thick loam cover is only present on the western part of the steepest slope section (fig.1). Elsewhere, the Tertiary deposits are very near to the surface. These deposits, which consist essentially of medium textured Brusselian sands are on the upper half of the field often covered with a sandy-loamy or even sandy-clayey material. On top of these, gravel is found, corresponding with the base of the Quaternary. The basal concavity consists of a thick colluvial deposit. The variable constitution of the soil profile is also reflected in the textural characteristics of the topsoil, as is demonstrated by the clay content of the topsoil, which is highly variable (fig.2).

The observations described in this paper were made from 15/11/83 to 03/10/84. At the start of the observation period the field was conventionally tilled and a seedbed was prepared. During the observation period, it was kept clear of weeds by the application of herbicides. At the foot of the slope a collector and a long-throated flume were installed, which was connected with a tape punch recorder, recording the water level every 5 minutes.

The evolution of the rill and gully pattern was monitored by periodic cartographic surveys, during which the po-

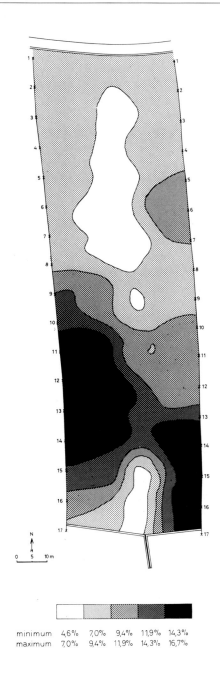

Figure 2: *Spatial variability of the clay content of the topsoil.*

sition of each rill, its mean width and maximum depth were noted on 15 transects every 10 meters downslope (fig.5 and 6). As most rills had a rectangular or trapezoidal cross-section, measurement of depth and mean width were sufficient to calculate the cross-sectional area. In most cases, measurements every 10 metres yielded sufficiently accurate information. When rill shapes were very irregular more measurements were made. Besides the rills, also flowlines, paths where there had been a concentrated flow of water without noticeable incision, were mapped. As bulk density of the topsoil was also regularly measured (GOVERS & POESEN 1986), the volumetric data could be used to calculate soil loss in the rill system. The only disadvantage of the technique is that, when an infill of the rill beds occurs due to the sediment produced on the rill sidewalls, there is an underestimation of the amount of hydraulic rill erosion.

3 GENERAL EVOLUTION

The total amount of rill and gully erosion on the field in a measuring period of less then one year (15/11/83–03/10/84), during which about 680 mm of rain with only moderate erosivity fell, was about 130 tons (fig.3), which illustrates the high erodibility of these loamy and sandy loamy soils.

Almost all (91%) of the rill and gully erosion occured in the periods 30/10/84–13/02/84 and 25/08/84–03/10/84 (fig.3). Several types of processes contributed to sediment production in the rills. A fundamental distinction has to be made between the deepening of the rills, caused by flowing water, and subsequent rill widening, caused by mass movement on

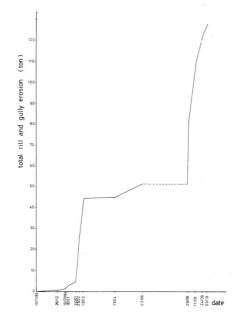

Figure 3: *Evolution of total rill and gully erosion in time.*

the sidewalls (phot.1). Rill widening was especially important during the period 29/08/84–03/10/84. On the Huldenberg field, the conditions for mass movement were clearly most favourable on the midslope section (transects 9–11). Therefore the total cross-sectional area of the rills was maximal at transect 10 on 03/10/84 (fig.4).

The peak erosion rates at transect 14 were caused by the occurrence of gullies on the eastern part of the steepest slope (phot.2, fig.6). The formation of these gullies was related to these areas where loose unaltered Brusselian sand was found directly below the sandy loamy plow layer. Sediment production in these gullies involved both hydraulic erosion and mass movement.

The part of the cross section of a rill or gully due to hydraulic erosion can

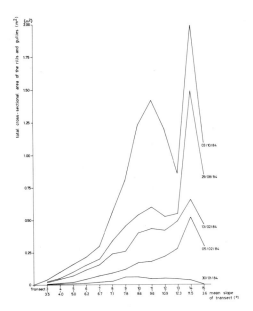

Figure 4: *Total cross-sectional area of the rills and gullies per transect at different dates. Transects are numbered consecutively from the upper to the lower border of the field.*

be calculated as the product of the bed width and the depth of the rill (BLONG et al. 1982). A number of measurements of the bed width in major rills (those most affected by sidewall processes) indicated that they had a mean width of 12 cm. It was assumed that in rills with a mean width greater than 12 cm the contribution of hydraulic erosion was equal to the depth multiplied by 12 cm. The rest of the cross section was attributed to mass movement processes. Sidewall processes in rills with a mean width smaller than 12 cm were neglected.

The upslope limit of a gully was formed by a clear headcut (phot.3). It was therefore easy to calculate separately the contribution of the gullies to the total sediment production. Where these gullies were located, a normal rill would have been formed if the bed had been cohesive as elsewhere on the field. The total amount of gully erosion was therefore diminished by the estimated amount of hydraulic rill erosion that would have occured when the subsoil would have

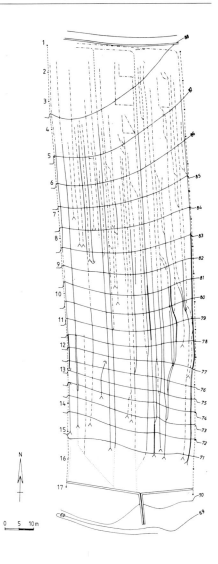

Figure 5: *Map of the rill and gully system (30/01/86, legend see fig.6).*

Photo 1: *Intense mass movement on rill sidewalls on the eastern part of the mid-slope section.*

Photo 2: *View on a gully developed in Brusselian sand on the western part of the steepest slope.*

been cohesive. The estimated amount of hydraulic erosion was added to the measured hydraulic rill erosion.

Apart from the surficial erosion due to concentrated water, it could be observed that on the loamy part of the field pipes were formed in which a considerable amount of water and sediment were transported (phot.4). Some attention had therefore to be given to this phenomenon.

Photo 3: *Headcut of a gully.*

4 HYDRAULIC RILL EROSION

4.1 TEMPORAL EVOLUTION

Total hydraulic rill erosion was about 67 tons (tab.1), most of which occured during three major runoff events, which had peak discharges between 70 and 90 l/s while the runoff rate during other runoff events was always below 25 l/s. Two of these events occured in the beginning of February (03/02/84 and 06/02/84) and were caused by very limited amounts of rain (8 and 13 mm, with a maximum intensity of about 50 mm/h). At that time, the soil was saturated, so that a runoff coefficient of nearly 100% was reached. Peak discharges and erosion amounts caused by these rainfall events were comparable to those caused by a 36 mm rainfall with a maximum intensity of 80 mm/h, which fell in summer (25/08/84) on a much drier soil. The first two rainfall events have a return period of 0.5-1 year for Middle Belgium, while the last one has a return period of about 50 years (LAURANT 1976). The fact that all three events caused similar amounts of hydraulic rill erosion (ca. 15 tons) illustrates the pronounced effect of antecedent moisture conditions on the erosive power of a given rainfall event.

4.2 THE INFLUENCE OF TOPOGRAPHIC FACTORS

On fig.7 the relation between the hydraulic rill intensity and the mean slope of the transect (expressed as a sine) is presented. The critical slope angle for rill initiation on these loamy soils is about 0.04-0.05 (ca.3 degrees): this agrees well with SAVAT & DE PLOEY's literature review (1982) and is related to the fact that the minimum threshold shear velocity for rill generation on these materials is about 3-3.5 cm/s (GOVERS 1985).

On 30/01/84 hydraulic rill erosion on the steepest slope section (transect 12, 13 and 14) is less than expected. This is due to the spatial variability in runoff generation. Visual observations showed that after tilling of the soil runoff generation started first on the upper and the basal part of the slope. Possible reasons for this are discussed in GOVERS & POESEN (1985). The runoff generated on the upper slope section caused relatively intense rill erosion on the mid-slope section (transects 7-10), but the water coming from upslope partly infiltrated on the steepest slope section so that some rills actually stopped here, depositing sediment on the steepest slope, or changed into pipes (fig.5).

Further hydraulic rill erosion occured mainly during major runoff events with general runoff generation. Hydraulic erosion was then most intense on the steepest slope section (fig.7). Nevertheless, the relationship between hydraulic rill erosion intensity and topographical factors remained variable. On 06/02/84 (after the first major runoff event), as is the case on 30/01/84, the relationship between hydraulic rill erosion intensity and slope is convex for the transects 3-10. Later the relationship becomes more or less linear. The relative amount of erosion on transect 15 is higher in the first half of the observation period than in the second. Further deepening of the rills on this transect was prevented by the proximity to the erosion base level. When slope steepness and slope length are considered simultaneously, using multiple (logarithmic) regression, this results in a decrease in time of the length exponent, together with an increase of the slope exponent (tab.2).

Date	Hydr. Rill Erosion	Mass Movement	Gully Erosion	Total
30/11/83	4.4	0.3	0.0	4.7
05/02/84	21.5	2.7	2.3	26.5
13/02/84	34.9	5.4	4.4	44.7
29/08/84	51.8	18.4	11.6	81.8
03/10/84	65.8	47.9	16.7	130.5

Table 1: *Sediment production by hydraulic rill erosion, mass movement on rill sidewalls and gully erosion on different dates.*

Photo 4: *Pipes due to mole activity partially turned into rills by collapsing of the roof.*

Date	Slope exponent	Length exponent	r^2
30/01/84	0.45	2.00	0.956
05/02/84	0.26	2.09	0.987
13/02/84	0.29	1.34	0.984
29/08/84	0.45	1.21	0.957
03/10/84	0.46	1.06	0.963

Table 2: *Slope and length exponents obtained by multiple (logarithmic) regression, expressing steepness and length.*

Regression equation

$$\sum A = A \log(S - S_{(cr)}) + B \log L + C \quad (1)$$

with
- $\sum A$ = the cross sectional area of the rills on a transect attributed to hydraulic erosion (m²)
- S = the mean slope of the transect (expressed as a sine)
- S_{cr} = the critical slope angle for rill initiation (set equal to 0.04)
- L = the distance from the upslope boundary of the field to the transect
- A, B, C = regression coefficients

The observations suggest that use of a relationship with constant slope and length exponents will not permit adequate description of temporal rill evolution. The fact that the relationship between the amount of hydraulic rill erosion, slope length and slope degree is variable was also found by MC COOL & GEORGE (1983), who found different slope and length exponents when rill systems measured on a number of fields in different years were considered. Serious problems can therefore also arise when fields with a different and complex topography have to be compared.

The variability of slope and length exponents is well-known (e.g. WISCHMEIER et al. 1958). MEYER et al. (1975a), FOSTER & MEYER (1975) and FOSTER et al. (1977) state that the variability of slope and length exponents observed during erosion plot research may reflect consideration of rill

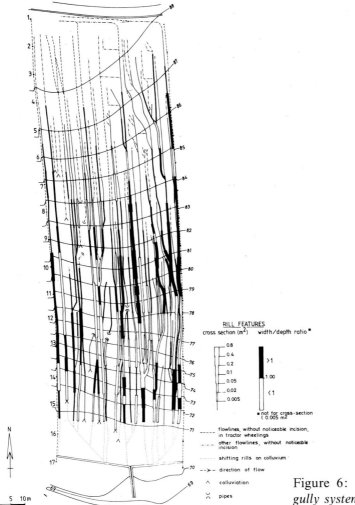

Figure 6: *Map of the rill and gully system (03/10/86).*

and interrill erosion together. Better results should therefore be obtained when both types of erosion processes are considered separately. However, even when hydraulic rill erosion alone is taken into account, a variation of slope and length dependency can occur. This variability seems also to exist for interrill erosion processes (BRYAN 1979, SINGER & BLACKARD 1982, VERHAEGEN 1984). Further research is therefore needed to clarify the reasons for this variability.

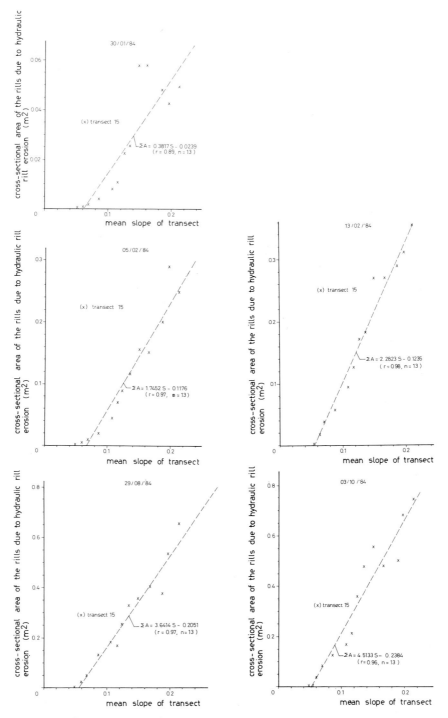

Figure 7: *Total cross-sectional area of the rills on a transect due to hydraulic rill erosion in function of the mean slope (expressed as a sine) of the transect.*

5 MASS MOVEMENT PROCESSES ON RILL SIDEWALLS

At the end of the observation period, about 37% (48 tons) of total erosion in the rill and gully-system originated in mass movement on rill sidewalls (tab.1). This figure demonstrates that these processes can contribute significantly to total sediment production in a rill system. Nevertheless, mass wasting is only occasionally mentioned in the literature in relation to rill erosion (e.g. MEYER et al. 1975b, VAN LIEW & SAXTON 1983) and in erosion modelling, rill erosion is considered to be only dependent of the erosive power of the flowing water (e.g. FOSTER 1982).

Most of the rill widening occured during the second half of the observation period, mainly during the very moist September (136 mm of precipitation) occuring after the major event of 25/08/84 (tab.1). It can therefore be seen as a postponed response of the rill system to the changes caused by an extreme event (overdeepening of the rills), allowing the system to adapt to a changed situation. The total impact of an extreme event is therefore even more important than one would expect from an estimation of the erosive power of the flow, as this is only related to hydraulic (vertical) erosion.

Mass movement processes were most active on the eastern half of the mid-slope section, resulting in large, relatively wide rills in this zone (fig. 6). Rill wall recession on the other parts of the field was less important.

From visual observations, it was obvious that in some cases undercutting triggered sidewall collapse. However, undercutting was not always a necessary condition for mass wasting to occur. This was shown by observations in a trough with subvertical walls. The trough was dug on 02/10/84, at the beginning of a very wet period. After 14 days, a niche was formed on the upslope side of the trough, at the contact of the plow layer and the subsoil, which enlarged and reached a maximum depth of about 15 cm (phot.5). Ultimately, the wall collapsed (phot.6). This niche can be interpreted as a miniature seepage cave caused by a lateral water flux at the contact between the plow layer and the less permeable subsoil. Periodic saturation of the plow layer was demonstrated by tensiometer recordings. Soil saturation may not only have caused wall instability due to seepage, but also due to collapsing on wetting of the soil at the contact between the plow layer and the subsoil, putting the plow layer under tension (BRADFORD et al. 1978). Basal saturation of the wall may also have been caused by the water flowing in the rill. All these factors may have destabilised rill sidewalls.

The fact that only on the eastern part of the mid-slope section the rill sidewalls were so severely affected by mass movement processes is due to the high sand content of the topsoil making it very prone to liquefaction upon saturation. The better structural characteristics of the plow layer inhibited deep-seated failures on the loamy part of the field, despite the presence of deep rills and periodic saturation of the plow layer. Rill wall recession occurred more slowly, partly by more or less gradual disaggregation of the topsoil material from the moment it became exposed to the open air, which can be seen as a kind of soil fall (IMESON & JUNGERIUS 1977), and partly by shallow sliding. The surficial sliding was related to wet conditions, while soil disaggregation was more or

Photo 5: *Formation of a niche on the upslope face of a trough on the eastern part of the mid-slope section (07/11/84).*

Photo 6: *Failure of the trough side as a result of niche formation (23/11/84).*

less continuous in time, with peak intensities during dry springtime, when drying of the topsoil caused intense crack formation on this part of the field (phot.7).

6 GULLY EROSION

Deep gullies developed on the eastern part of the steepest slope section where unweathered Brusselian sands were situated very near the surface. They produced 16.7 tons of sediment, although the three gully systems had a total length of only 45 m, while the whole rill and streamline pattern on the field had a length of about 2200 m (tab.2).

The typical evolution of these gullies can be illustrated by taking the gully formed in the eastern tractor wheel tracks as an example (fig.8). The gully was initially formed by headcut-erosion, which started at the place where the rill bottom first reached the sand surface and subsequently migrated upslope. However, upslope migration of the headcut and deepening of the bed occurred only during major runoff events: during the runoff event of 25/08/84 the headcut of the gully migrated 6 m upslope and the gully was deepened by more than 50 cm over the greatest part of its length. During periods without major runoff vertical and headcut-erosion were virtually absent and sidewall collapse caused the gully to widen.

The evolution of the gully sidewalls differed considerably from that of the rill sidewalls, located in cohesive materials. The erosion resistance of the sand

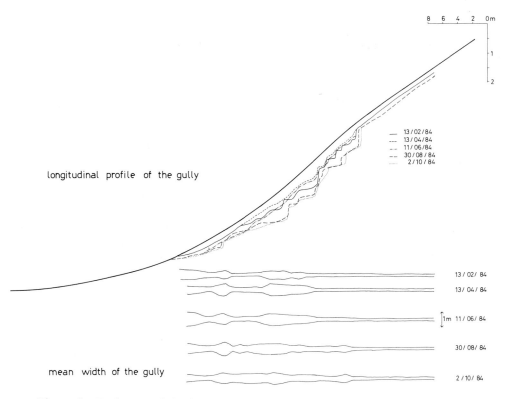

Figure 8: *Evolution of the longitudinal profile and the mean width of a gully.*

was indeed smaller than that of the topsoil, so that the bed width was often much greater than the top width. The tension on the sidewalls caused the formation of cracks and consequent failure of the wall. During the following rainfall events, the loose sediment was evacuated (with occasional formation of short pipes underneath the collapsed material), resulting in a more or less rectangular cross-section. Due to the widening of the gully, the walls became exposed to sunlight. As the sand became dry, it lost its cohesion and an equilibrium slope was formed (phot.8). The loose sand was again evacuated during the next important runoff event.

Contrary to the rill sidewalls, the gully sidewalls were most active during dry periods, when the sand lost all its consistency. Undercutting was the primary cause of sidewall instability, while (periodic) saturation of the plow layer did not occur, because the sand allowed a sufficient vertical drainage.

7 SOIL PIPING

From the beginning of spring, pipeflow and erosion could be observed in numerous pipes in the loamy part of the field. The original formation of the pipes was in most cases caused by mole activity, which was, due to better moisture conditions, most intense on this part on the field. Once the pipes were formed,

Photo 7: *Soil fall on rill sidewalls on the western part of the steepest slope.*

Photo 8: *Equilibrium slope formed on a gully sidewall due to dry sliding of the sand.*

a considerable amount of pipeflow and erosion could take place: lateral flow was promoted by the low permeability of the compacted subsoil (<1 mm/h). Pipe formation could also occur without animal interference: pipes were observed to develop under longitudinal microdepressions caused by the tilling of the topsoil. As these depressions were continuous from the top of the slope to the base, they concentrated considerable amounts of runoff water on the upper part of the slope. When this water infiltrated on the steepest slope section, this caused an increased throughflow, which could, maybe in combination with differences in soil bulk density, give rise to the formation of a soil pipe (phot.9).

Also on these cultivated soils an important subsurface movement of water and sediment is possible. Necessary conditions for this are an impervious subsoil overlain by a more permeable plow layer with sufficient structural characteristics, so that the pipes don't collapse immediately after formation, a sufficient slope gradient and some process causing a linear discontinuity of the bulk density and the macroporosity of the topsoil.

It was not possible to measure directly the amount of soil that was eroded in the pipes. Only when the pipe turned into a rill (because the roof collapsed), could the rill volume be measured. Besides the formation of pipes, mole activity also contributed directly to sediment erosion by supplying loose sediment to existing rills. The importance of this sediment production could be estimated by collecting the loose sediment in the rills over a length of 1 meter just upslope from each transect.

Photo 9: *Pipe formed on a linear discontinuity due to the tilling of the soil.*

The collected amount was then multiplied by 10. During a period of about two and a half months in springtime (01/03/84–16/05/84), when mole activity was at a maximum, moles produced in this way about 750 kg of sediment, which were evacuated during consequent runoff events. Taking into account that nearly all of this sediment was produced on the loamy part of the field, with a surface area of about 0.25 ha, this quantity corresponds with a sediment loss of 3 tons/ha. The amount of sediment eroded in the mole pipes was certainly much higher, so that it may be stated that animal activity on arable land can directly or indirectly cause erosion losses of more than 10 tons/ha. Although animal activity may not be as important in an agricultural environment as in less disturbed forest or desert environments (IMESON & KWAAD 1976, YAIR & LAVEE 1981), its influence on erosional and hydrological processes may certainly not be neglected.

8 CONCLUSIONS

A periodically-carried out cartographic survey can yield basic information concerning the development of a rill system in time and in space. This kind of technique, which was, until present, only occasionally used in soil erosion research, does not require any stationary measuring installation, which make plot measurements often very expensive and time-consuming. The major advantage of this method is that it allows an insight in the dynamics of an erosional system as it develops on a whole slope. Plot measurements often yield only information about total soil loss on a limited slope segment, so that vital information concerning several factors with a major influence on rill development will never be acquired.

The study of the evolution of a rill system during one year made it clear that it is not possible to understand (or model) rill erosion dynamics without this information which allows a distinction between the different erosion processes and forms that are active.

Hydraulic rill erosion is, when runoff generation is general, topographically controlled. However, the relationship between hydraulic rill erosion intensity, slope steepness and slope length is variable in time, which might cause specific problems with respect to the description of temporal rill evolution and the comparability of measurements on different sites.

The evolution of the rill sidewalls is not only dependent on rill depth but mainly on the mechanical and the hydrodynamical properties of the different soil layers. Sediment production on rill sidewalls is therefore not necessarily maximal on the steepest slope. When saturation of a plow layer with bad structural characteristics occurs, mass movement can affect rill sidewalls over a depth of several decimeters and may become very important in total sediment production. When the soil material has a better structure, rill walls recede dominantly by soil fall (favoured by crack formation) and shallow sliding.

Gully erosion occurs when the subsoil consists of loose sands and has its own dynamics. The regression of the gully headcut and the deepening of the bed are limited to major runoff events. Sidewall collapse is strongly related to undercutting by the flow. In contrast with the loamy, cohesive rill sidewalls, the sandy gully sidewalls evolve most rapidly in dry periods.

The occurrence of these gullies poses a specific problem when the erodibility of a given soil has to be considered. To evaluate the risk of such a disastrous erosion form, not only knowledge of the specific circumstances at the location where gully formation may take place is required: it is for instance clear that the formation of gullies is not only determined by the constitution of the soil profile, but was also due to an important water supply from upslope. The vulnerability of such a soil profile is therefore dependent on its position in the landscape: no unique erodibility-value can therefore be attributed to such a soil profile.

Another striking observation is the fact that also on cultivated loamy soils an important amount of pipeflow and - erosion can occur, which is in most cases initiated by animal activity. The biotic factor may not, therefore, be neglected with respect to soil erosion on arable land.

Many hydrologic and erosion processes on arable land are thus not only related to the characteristics of the topsoil, but also to the hydrological and mechanical properties of the subsoil. This has important implications for e.g. erosion risk mapping.

ACKNOWLEDGEMENT

Many thanks are indebted to prof. De Ploey for his help during various stages of this work, to prof. F. Depuydt, C. Steenmans and F. Roselle, who made a topographical map of the field, R. Geeraerts and R. Witters who made the drawings, ir. M.G. Bos (ILRI, Wageningen), who designed the measuring flume, G. Wyseure, who wrote a computer program allowing to analyse the rainfall and runoff data and Dr. Paulissen, whose work on the spatial variability of the surficial deposits of the field was very helpful to me for the interpretation of my data and who assisted also in the fieldwork for the preparation of the topographical map.

REFERENCES

[BLONG et al. 1982] BLONG, R., GRAHAM, O. & VENESS, J.: The role of sidewall processes in gully development: some N.S.W. examples. Earth Surface Processes 7, 381–385.

[BOARDMAN 1984] BOARDMAN, J.: A morphometric approach to soil erosion on agricultural land near Lewes, East Sussex. In: Lakehurst, C. & Grant, R. (ed.), Issues in Countryside Research, Kingston Brighton Research Papers, Brighton Polytechnic, 1–10.

[BOLLINNE 1977] BOLLINNE, A.: La vitesse de l'erosion sous culture en region limoneuse. Pedologie, 27, 191–207.

[BOLLINNE 1982] BOLLINNE, A.: Etude de prevision de l'erosion des sols limoneux en

Moyenne Belgique. Unp. Ph.D. thesis, Universite de Liege, Faculte des Sciences, 365 p.

[BRADFORD et al. 1978] BRADFORD, J., PIEST, R. & SPOMER, R.: Failure sequence of gully headwalls in western Iowa. Soil Sci. Soc. Am. J., **42**, 323–327.

[BRYAN 1979] BRYAN, R.B.: The influence of slope angle on soil detachment by sheetwash and rainsplash. Earth Surface Processes, **4**, 43–58.

[FOSTER 1982] FOSTER, G.: Modeling the erosion process. In: Haan, C. (ed.), Hydrologic Modeling of Small Watersheds. ASAE, St. Joseph, 297–370.

[FOSTER & MEYER 1975] FOSTER, G. & MEYER, L.: Mathematical simulation of upland erosion by fundamental erosion mechanics. In: Present and Prospective Technology for Predicting Sediment Yields and Sources. Agr. Res. Service Rep. ARS-S-40, 190–207.

[FOSTER et al. 1977] FOSTER, G., MEYER, L. & ONSTAD, C.: An erosion equation derived from basic erosion principles. Trans. ASAE, **20**, 678–682.

[FULLEN 1985] FULLEN, M.: Compaction, hydrological processes and soil erosion on loamy sands in East Shrophire, England. Soil and Tillage Research, **6**, 17–29.

[GABRIELS et al. 1977] GABRIELS, D., PAUWELS & DE BOODT, M.: A quantitative rill erosion study on a loamy sand in the hilly region of Flanders. Earth Surface Processes, **2**, 257–259.

[GOVERS 1985] GOVERS, G.: Selectivity and transport capacity of thin flows in relation to rill erosion. CATENA **12**, 35–49.

[GOVERS & POESEN 1985] GOVERS, G. & POESEN, J.: A field-scale study of soil sealing and compaction on loamy and sandy loam soils. Part I. Spatial variability of soil surface sealing and compaction. In: Callebaut, F., Gabriels, D. De Boodt, M. (ed.): Assessment of Soil Sealing and Crusting. Proceedings of the Symposium held in Gent, 1985, 171–182.

[IMESON & JUNGERIUS 1977] IMESON, A. & JUNGERIUS, P.: The widening of valley incisions by soil fall in a forested Keuper area, Luxembourg. Earth Surface Processes, **2**, 141–152.

[IMESON & KWAAD 1976] IMESON, A.C. & KWAAD, F.J.M.P.: Some effects of burrowing animals on slope processes in the Luxembourg Ardennes. Part 2: The erosion of animal mounds by splash under forest. Geografiska Annaler, **58 Ser. A**, 317–328.

[LAURANT 1976] LAURANT, A.: Nouvelles recherches sur les intensites maximums des precipitations a Uccle: courbes d'intensite-duree-frequence. Annales de Travaux Publics de Belgique, **4**, 320–328,

[MC COOL & GEORGE 1983] MC COOL, D. & GEORGE, D.: A second-generation adaption of the Universal Soil Loss Equation for pacific northwest drylands. ASAE paper 83-2066, 20 p.

[MEYER et al. 1975a] MEYER, L., FOSTER, G. & ROMKENS, M.: Source of soil eroded by water eroded from upland slopes. In: Present and Prospective Technology for Predicting Sediment Yields and Sources, Agr. Res. Service Rap. ARS-S-40, 177–189.

[MEYER et al. 1975b] MEYER, L., FOSTER, G. & NICHOLOV, S.: Effect of flow rate and canopy on rill erosion. Trans. ASAE, **18**, 905–911.

[SAVAT & DE PLOEY 1982] SAVAT, J. & DE PLOEY, J.: Sheetwash and rill development by surface flow. In. Bryan, R. & Yair, A. (ed.), Badland Geomorphology and Piping. Geo Books, Norwich, 113–126.

[SINGER & BLACKARD 1982] SINGER, M. & BLACKARD, J.: Slope angle-interrill soil loss relationships for slopes up to 50%. Soil Sci. Soc. Am. J., **46**, 1270–1273.

[VAN LIEW & SAXTON 1983] VAN LIEW, M. & SAXTON, K.: Slope steepness and incorporated residue effects on rill erosion. Trans. ASAE, **26**, 1738–1744.

[VERHAEGEN 1984] VERHAEGEN, T.: Laboexperimenten en terreinwaarnemingen in verband met de erosiegevoeligheid van lehmige bodems. Unp. Ph. D. thesis, 213 p.

[WISCHMEIER et al. 1958] WISCHMEIER, W., SMITH, D. & UHLAND, R.: Evaluation of factors in the soil loss equation. Agricultural Engineering, **39**, 458–462.

[YAIR & LAVEE 1981] YAIR, A. & LAVEE, H.: An investigation of source areas of sediment and sediment transport by overland flow along arid hillslopes. Erosion and Sediment Transport Measurement, Proc. of the Florence Symposium, IAHS Publ. **133**, 433–466.

Address of author:
Gerard Govers
Senior Research Assistant, National Fund for Scientific Research, Belgium
Laboratory of Experimental Geomorphology
Catholic University of Leuven
Redingenstraat 16B
3000 Leuven, Belgium

TRANSPORT OF ROCK FRAGMENTS BY RILL FLOW—A FIELD STUDY

J. **Poesen**, Leuven

SUMMARY

Field measurements on a field plot in the Belgian loam region have been performed to find out 1) which erosion processes are capable of moving rock fragments at the surface of stony soils, 2) under which conditions do rock fragments start moving on hillslopes, 3) which factors determine displacement distances and, 4) how intense rock fragment transport can be on upland areas. Monitoring coloured rock fragments, placed on interrills as well as in rills, revealed that during a moderate rainfall event rock fragments up to 9.0 cm in diameter travelled downslope by rill flow. The competence of interrill flow was about one order of magnitude smaller. Furthermore, it was found that conditions of incipient motion for single or clustered rock fragments, lying on a rill bed, correspond to a SHIELDS' criterion value (Θ_c) of 0.015 rather than 0.05. Rock fragment transport distance was controlled more by fragment size then by fragment shape. With respect to rill site characteristics, rock fragment transport distance correlated better with rill slope than with peak rill flow discharge and, hence, also with rill catchment area. The product of stream power and mean specific potential energy of a rock fragment at a given rill site explained 61 per cent of the variation in rock fragment transport distance between the rill sites. Estimated maximum intensity of rock fragment transport on the field plot by rill flow equals 256 kg/m/year, a figure exceeding by two orders of magnitude maximum sediment transport rate by raindrop splash. Implications of these results with respect to a soil conservation technique, the soil erodibility factor (K) of the USLE, rill channel armoring, taphonomy, the genesis of stony colluvium and stream channel response to rock fragment transport on uplands by rill and gully flow are further discussed.

RESUME

Des mesures effectuées sur un versant de la région limoneuse en Belgique avaient pour but de préciser (a) les processus d'érosion susceptibles de transporter des éléments grossiers à la surface de sols caillouteux, (b) les conditions hydrauliques nécessaires à leur mobilisation, (c) les facteurs déterminants la distance des déplacements et (d) le volume susceptible de migrer ainsi le long des versants. Réalisé dans ce but, le suivi des déplacements de graviers et cailloux peints de différentes tailles, placés dans des rigoles et sur les zones inter-rigoles a permis de montrer que, lors d'une pluie

ISSN 0722-0723
ISBN 3-923381-07-7
©1987 by CATENA-Verlag,
D–3302 Cremlingen-Destedt, W. Germany
3-923381-07-7/87/5011851/US$ 2.00 + 0.25

d'intensité modérée, le diamètre maximum des éléments mobilisés dans les rigoles était de 9 cm, tandis que sur les inter-rigoles il était dix fois moindre. En outre, les résultats indiquent que, pour mobiliser des éléments isolés ou groupés posés sur le fond des rigoles, le critère de SHIELDS est égal à 0.015 au lieu de 0.050, valeur admise en général. La longueur du déplacement des éléments grossiers semble contrôlée davantage par leur taille que par leur forme. D'autre part, la corrélation de la distance de transport avec le gradient de pente de la rigole s'est révélée meilleure que celle avec le débit unitaire maximum (établi pour le bassin de la rigole). En chaque site, le produit de la puissance du courant et de l'énergie potentielle spécifique moyenne expliquait 61% de la variation entre les différentes longueurs des déplacements. Pour l'ensemble du versant, le débit maximum d'éléments grossiers transportés dans les rigoles a été évalué à 256 kg/m/an; par comparaison, ce débit est supérieur de deux ordres de grandeurs au débit solide maximum dû au splash. Les implications de ces résultats peuvent être importantes, en particulier pour:

- l'utilisation des cailloux dans la protection des sols,
- l'érodibilité des sols (facteur K de la USLE),
- la formation de pavages dans le fond des rigoles,
- l'archéologie,
- la genèse des colluvions caillouteux,
- la dynamique fluviale.

1 INTRODUCTION

In the past, erosion processes by water have been mainly studied by soil scientists and agriculturists on rock-free soils, essentially because of their relative crop production. Yet, soils containing rock fragments, i.e. particles 2 mm or larger in diameter including all sizes that have horizontal dimensions less than the size of a pedon (MILLER & GUTHRIE 1984), in their surface, represent a significant portion of our land resources to be increasingly used for food and fiber production with the ever present potential of undesirable effects of wind and water erosion (NIELSEN 1984). Regardless of the food and fiber issue, effective erosion control of these soils will always be an important concern. Information on the erodibility behaviour of soils containing rock fragments is especially needed because of their number, their extent and their potential benefits or limitations for landuse (NICHOLS et al. 1984). Before existing erosion-estimating models such as the Universal Soil Loss Equation (WISCHMEIER & SMITH 1978) can be applied to these soils, more information on the kind and intensity of erosion processes, acting on these soils is needed (SIMANTON et al. 1984).

It is generally accepted that large contents of rock fragments in and on the soil surface reduce interrill and rill erosion.

Furthermore, BOX & MEYER (1984, p.86) state that: "... the erodible parts of the land surface are the soil materials "(less than 2 mm)", and not coarse fragments. The coarse fragments generally remain even when serious soil loss is occurring.". Field observations on agricultural lands in Mid-Belgium, however, reveal that considerable amounts of rock fragments are transported from upland

Photo 1: *Recent colluvial deposits containing considerable amounts of rock fragments, originating from the Quaternary base gravel deposits which outcrop on eroded upslope sections or which are ploughed up on uplands by farming (4.4.1985, Huldenberg, Belgium).*

Photo 2: *Deposited rock fragments along a rill channel. Length of stick equals 23 cm (4.4.1985, Huldenberg, Belgium).*

areas and deposited at the foot of slopes (photos 1, 2). Hence, the objectives of this field study were to answer the following questions:

1. Which erosion processes are capable of moving rock fragments at the surface of stony soils?

2. Under which conditions do rock fragments start moving on slopes?

3. Which factors determine displacement distances of rock fragments?

4. How intense is rock fragment movement on slopes?

2 MATERIALS AND METHODS

The field measurements were carried out on the 170 m long and 45 m wide Huldenberg field plot, situated in central Belgium and described in more detail by GOVERS & POESEN (1986). Texture of the top soil varies from silt loam and loam to sandy loam. Due to the occurrence of gravel rich fluviatile deposits—probably of early Quaternary age (PAULISSEN, pers. comm.)—at shallow depth, soils can locally have up to 25% by volume of gravel and occasional cobbles in the surface 10 cm. So, at these sites soils can be termed gravelly/cobbly loam or gravelly/cobbly sandy loam. Rock fragments consists mainly of rounded to well rounded flint pebbles.

After the field plot was placed in a conventional seedbed on November 15th 1983, the soil surface was kept bare. As a consequence, surface sealing, compaction of the plow layer and rilling occurred in the next year (POESEN & GOVERS 1986). On November 7th 1984, 5 interrill sites and 4 rills were selected on the basis of the slope of the interrill site and the dimensions of the rill cross section (fig.1). Through each interrill site a transect, more or less parallel to a contour, was also selected. Each transect was named after the colour of the painted rock fragments used. Following this, 687 selected flint pebbles and cobbles, with intermediate diameters ranging between 0.35 and 9.8 cm and with variable shape (flatness index (F.I.) varied between 1.1 and 3.9, $F.I. = (L + I)/(2S)$ with L denoting the longest, I the intermediate and S the shortest dimension of the fragment along three perpendicular axes) and variable roundness, were coloured and numbered. Close to almost each crossing of a transect with one of the 4 selected rills, three places, about 1 m apart, were selected on the rill bed. After measuring bottom width and slope of the rill bed (tab.1), at each place a set of 5 to 35 coloured rock fragments with variable size, shape and roundness was placed randomly in a cluster on the rill bed. At the same time, coloured fine pebbles with intermediate diameters ranging between 0.2 and 0.8 cm and with variable flatness were placed on a 1 m long line, parallel to a contour, at each selected interrill site. On November 26th 1984, after a rainy period, the coloured rock fragments on the interrills and in the rills were recovered and their transport distances were measured. From the 687 coloured fragments used as a tracer in the rills, 71% were found again. After mapping flowlines, visible on the interrills, on a topographic map with scale 1:200, catchment area of each rill upslope of the place where coloured fragments were laid down, was calculated (tab.1).

A recording rain gauge, installed at a distance of 650 m from the field plot, provided the necessary rainfall data.

3 RESULTS AND DISCUSSIONS

3.1 COMPETENCE OF INTERRILL AND RILL FLOW

Rainfall data, recorded during the period of observation, are shown in tab.2. During the period from 14/11 to 21/11, the finest pebbles on the interrills moved a maximum downslope distance of 1 cm while in the rills rock fragments with an intermediate diameter up to 5.0 cm had moved. During the subsequent period,

Transport of Rock Fragments by Rill Flow

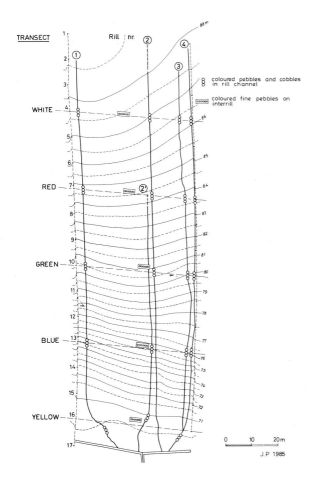

Figure 1: *Topographic map of the Huldenberg field plot and location of the selected interrill and rill sites.*

TRAN-SECT	RILL No	$\overline{S}$	$\overline{b}$ (cm)	A (m²)
White	1	0.049	5.0	2.4
White	2	0.063	13.3	16.8
White	3	0.062	12.0	7.0
White	4	0.082	10.0	30.2
Red	1	0.144	10.2	16.5
Red	2	0.122	16.3	28.2
Red	3	0.129	7.0	31.7
Red	4	0.125	8.7	32.7
Green	1	0.172	5.3	23.2
Green	2	0.134	15.0	32.1
Green	3	0.156	14.0	61.8
Green	4	0.189	13.0	45.1
Blue	1	0.202	7.0	28.9
Blue	2'	0.268	3.0	10.7
Blue	3	0.175	9.0	65.7
Blue	4	0.219	16.0	48.6
Yellow	1	0.074	48.0	159.2
Yellow	2'	0.075	52.3	12.9
Yellow	3&4	0.063	15.0	143.5

Table 1: *Some characteristics of the selected rill sites.*

$\overline{S}$ = mean gradient of rill beds ($n = 3$).
$\overline{b}$ = mean width of rill bottom ($n = 3$).
A = catchment area of rill, upslope of the selected rill site.

the largest pebbles on the interrills, i.e. fragments with an intermediate diameter of 0.8 cm, moved a maximum downslope distance of 5 cm while considerably larger particles moved in the rills. Tab.3 lists for each rill site the mean maximum intermediate diameter of rock fragment moved by rill flow. This value was calculated as the mean intermediate axis of the 1 to 5 largest rock fragments which were moved by the flow and the 1 to 5 smallest rock fragments which did not move. So, each value represents an intermediate diameter of a rock fragment which has only just become mobile at the peak stress value exerted by the flow on the rill bed. It is assumed that rock fragment movement could be attributed to the rainfall event causing largest peak discharges between dates of inspection. In our situation, peak discharge corresponded to a rainfall event with a maximum rainfall intensity of 30 mm/h during 12 min (tab.2). Such a rainfall event has a return period of 6 months in central Belgium (LAURANT 1976).

From the preceding, it is clear that the competence of the rill flow, i.e. the ability of the flow to transport rock fragments as measured by the size of the largest fragment it can move, exceeds almost by a factor 10 the competence of the interrill flow. On interrill areas we consider splash-creep (MOEYERSONS & DE PLOEY 1976) and runoff-creep (DE PLOEY & MOEYERSONS 1975) to be responsible for the slight downslope displacement of the fine pebbles. If one assumes a uniform distributed sheet flow on the interrill soil surface of the field plot, one can easily calculate that during peak runoff, sheet flow unit discharge certainly never exceeded 10 cm^2/s. As shown by laboratory experiments of DE PLOEY & MOEYERSONS (1975), such flow discharges are usually not competent to transport rock fragments with diameters ranging between 1 and 8 cm. On the other hand, rill flow caused by a moderate rainfall event is capable of moving rock fragments with a mean intermediate diameter of up to 9 cm (tab.3). Hence, rill flow and other forms of concentrated flow (e.g. ephemeral gully flow) can be held responsible as the most important processes which evacuate pebbles and cobbles from upland areas. Because of the minor importance of interrill flow with respect to the evacuation of rock fragments from slopes, most attention was further paid to rock movement by rill flow.

3.2 INCIPIENT MOTION CONDITIONS FOR ROCK FRAGMENTS BY RILL FLOW

Sediment threshold data for turbulent flow are usually expressed in terms of SHIELDS' model, in which the forces resisting particle motion are balanced against motivating forces. When motivating forces surpass resisting forces, the particle moves. SHIELDS produced non-dimensional relationships between the density of the sediment (ρ_s), the fluid density (ρ), the grain diameter (D), the kinematic fluid viscosity (v), the acceleration due to gravity (g) and the shear stress of the fluid flow (τ_o), such that (VANONI 1977, 96):

$$\Theta = \frac{\tau_o}{(\rho_s - \rho)gD} = f(u_* D/v) \qquad (1)$$

in which $u_* = (\tau_o/\rho)^{0.5}$ and f denotes function of. Θ is the SHIELDS entrainment function and needs to be assigned a critical value (Θ_c = SHIELDS' criterion) in order to solve the left hand term

PERIOD	volume (mm)	RAINFALL maximum rainfall intensity
07/11–14/11	0	0
14/11–21/11	19.5	4.25 mm/h during 24 min (= 1.7 mm)
21/11–26/11	43.5	30.0 mm/h during 12 min (= 6.0 mm)
07/11–26/11	63.0	30.0 mm/h during 12 min

Table 2: *Rainfall data recorded during the period of field observation.*

TRANSECT	RILL No			
	1	2	3	4
White	1.2	6.2	2.6	5.7
Red	4.3	7.8	5.4	9.0
Green	4.2	6.1	6.0	6.3
Blue	5.7	5.8	7.5	5.8
Yellow	4.6	5.4		7.0

Table 3: *Mean maximum intermediate diameter (cm) of rock fragments, moved by rill flow during the period of observation.*

of equation (1) for a given fragment diameter or stress value.

In order to compare our field data related to rock fragment initiation by rill flow, with existing theory, an attempt was made to calculate Θ. Since no detailed measurements of flow depth or flow velocity were made during peak flow, an accurate calculation of τ_o or u_* was impossible. Hence, we used the approach outlined by BETTES (1984). This author combined SHIELDS'criterion for critical flow (i.e. $\Theta_c = 0.05$) with the rough-turbulent friction law to provide a simple relationship between the critical discharge (q_c) and slope (S). For rock fragments with specific gravity 2.65 and water as the fluid, this relation becomes (fig. 2):

$$q_c = \frac{0.104}{S} log_{10}(\frac{1.221}{S}) \quad (2)$$

with

$$q_c = \frac{q}{\sqrt{gD^3(\rho_s/\rho - 1)}} \quad (3)$$

where
q = critical unit flow discharge.

For each rill site where rock fragments were laid down, q_c was calculated according to equation (3). Based on estimates of the velocities of overland flow

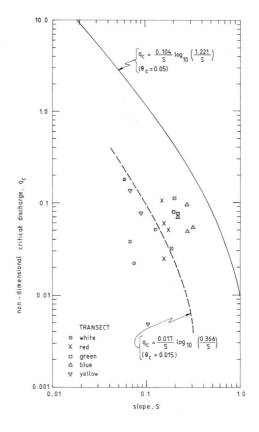

Figure 2: *Critical discharge (equation 3) plotted against rill bed gradient for Huldenberg rill data.*

and channel flow, it could be safely stated that the time of concentration of each rill catchment, i.e. the time required to reach a steady runoff state, was well below the duration of the rainfall event causing peak runoff (i.e. at least 12 min, tab.2). Hence peak rill flow discharge (Q) was calculated by the rational formula:

$$Q = CIA \qquad (4)$$

with
- Q = flow discharge (cm^3/s);
- C = runoff coefficient. At the moment of peak rainfall intensity, water content of the top soil exceeded field capacity. In addition, top soils were strongly sealed and compacted at that time (GOVERS & POESEN 1986). Hence, we assumed C to be equal to 1.0 in this study;
- I = rainfall intensity ($cm^3/s/m^2$). In our case, I equalled 8.33 $cm^3/s/m^2$ for a peak rainfall intensity of 30 mm/h (tab.2);
- A = rill catchment area (m^2, table 1).

Next, unit peak rill flow discharge (q) was calculated as Q/b_s where b_s represents the smallest rill bottom width for each rill site. The latter value was chosen out of three b-values for each rill site in order to obtain the maximum possible peak unit rill flow discharge. Calculated q-values at the Huldenberg rill sites varied between 2.4 and 99.7 cm^2/s. D, representing the intermediate diameter of an incipient moving rock fragment for a given rill site, was taken from tab.3 while $\rho s/\rho$ was set equal to 2.65 since flint fragments were used as a tracer.

A plot of calculated q_c-values versus maximum slope of the rill bed is shown in fig. 2. With the exception of two rill sites, i.e. rill 3 (section white) and rill 2' (section yellow), peak flow was turbulent, i.e. Reynolds number ($= q/v$) exceeded 500. From fig.2 it can be seen that q_c-data corresponding to incipient motion of the rock fragments in the Huldenberg rills are situated, in the mean, one order of magnitude below the q_c-values representing SHIELDS criterion. The discrepancy between q_c-values corresponding to our field data and q_c-values calculated with equation (2) is largest for the smallest rill bed gradient and decreases towards steeper slopes. The differences between both q_c-values as shown in fig. 2, are minimal since we used maximum possible q- and S-values. It is very likely that during peak flow, q-values were somewhat lower than those used for the calculation of q_c because the use of a lower runoff coefficient and a larger mean rill bottom width would have been more realistic. Consequently, it can be safely said that SHIELDS criterion ($\Theta_c = 0.05$) overpredicts, at least by one order of magnitude, real incipient motion conditions for loose rock fragments lying on a rill bed.

The discrepancy between our field observations and SHIELDS' model has several reasons. In order to develop his model, SHIELDS used results from almost equidimensional grains laid in flat beds. This was clearly not the case in our field measurements where a set of 5 to 35 rock fragments were laid on a relatively smooth rill bed. Hence, these fragments protruded considerably above the mean bed elevation. Experiments of FENTON & ABBOTT (1977) and field data compiled by ANDREWS (1983) clearly show that SHIELDS' criterion (Θ_c) decreases with an increasing degree of exposure of individual grains to the fluid flow. Furthermore, SHIELDS' criterion applies to conventional open channel flows with gentle slopes, large ratios of depth to sediment size and fine bed materials (sand and fine gravels). Recent studies (e.g. BATHURST et al. 1982) have shown that for flows with steep slopes, i.e. larger

TRAN-		RILL No		
SECT	1	2	3	4
White	-0.41*	-0.40*	-0.58***	-0.13
Red	-0.44**	-0.13	-0.56***	-0.41**
Green	-0.65***	-0.50	-0.31	-0.50***
Blue	-0.09	-0.30	-0.72***	-0.62***
Yellow	-0.34	-0.55**		-0.19

* significant at 0.10 level,
** significant at 0.05 level,
*** significant at 0.01 level.

Table 4: *Correlations between intermediate rock fragment diameter and transport distance for each rill site, as indicated by product moment correlation coefficient.*

than 0.005, with depths which are of the same order of magnitude as the bed material size and with coarse bed materials, this criterion does not apply because the processes of sediment entrainment are not adequately described by shear stress considerations. This too explains partly the observed discrepancy since rill beds are generally steeper than 0.05 and since rill flow depth is relatively small compared to fragment diameter. In addition, it has been observed in steep, narrow rill channels where flow depth is smaller than rock fragment size, e.g. 10 cm, that a fragment lying on the bed obstructs the flow and forms a small dam. As a consequence, a small water reservoir is built up at the upslope side of the fragment, creating an extra hydraulic pressure on the fragment which enhances fragment movement. The phenomenon is typical for narrow rill channels with relatively large rock fragments lying on their beds.

From the preceding, it can be concluded that our data are not adequate to investigate if a SHIELDS-type relation governs incipient motion conditions of single or clustered rock fragments in steep narrow channels or not. However, our data show that incipient motion conditions for these rock fragments are overestimated by the classical SHIELDS' criterion (i.e. $\Theta_c = 0.05$). Hence, when calculating threshold fragment sizes for bed load transport in such channels (e.g. THARP 1984), an appropriate Θ_c-value should be taken into account. Our field data suggest that a mean Θ_c-value of 0.015 is more realistic. This Θ_c-value, combined with the approach of BETTES, leads than to the equation (fig. 2):

$$q_c = \frac{0.0171}{S} log_{10}(\frac{0.366}{S}) \qquad (5)$$

3.3 FACTORS CONTROLLING DISPLACEMENT DISTANCES FOR ROCK FRAGMENTS IN RILLS

3.3.1 Rock Fragment Properties

Properties of rock fragments investigated in this study were size (intermediate diameter) and shape (F.I. = flatness index). For each rill site, intermediate rock fragment diameter was plotted versus corresponding displacement distance. Fig.3 illustrates how diameter of rock fragments relates to distance moved for some selected rill sites. Tab.4 depicts for each rill site correlation coefficients with respect to this relation.

From fig.3 it can be seen that displacement distances of rock fragments as a consequence of rill flow during a moderate rainfall event can be considerable: e.g. up to 84 m for a 3.6 cm diameter fragment (transect green, rill 3) or 2.7 m for a 9.8 cm diameter fragment (transect red, rill 2, not shown in fig. 3). In general, a negative relation between

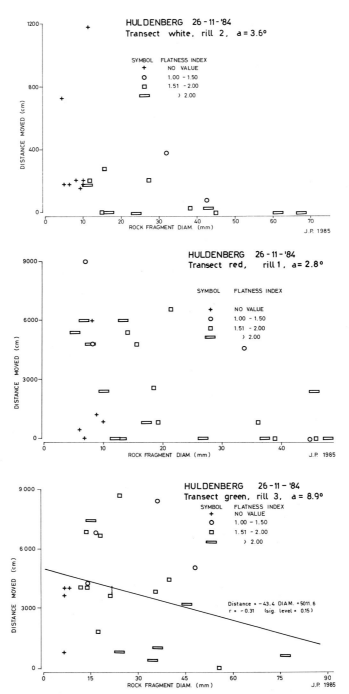

Figure 3: *(a–c) Relationship between rock fragment diameter and displacement distance in rills for some selected sites.*

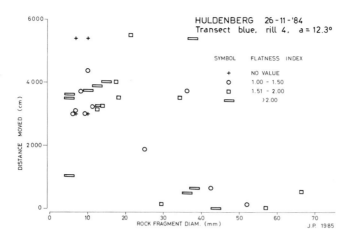

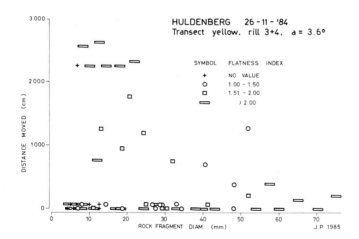

Figure 3: *(d–e) Relationship between rock fragment diameter and displacement distance in rills for some selected sites.*

fragment size and distance moved is observed. For a given fragment diameter, however, a large variation in transport distance exists. This variation decreases as size increases. The large scatter is also reflected by the low correlation coefficients (table 4). This is partly attributed to the stochastic nature of gravel entrainment by rill flow.

A visual inspection of all scatter diagrams leads to the conclusion that there is only a slight tendency, i.e. for 7 out of the 19 rill sites, for the round fragments (i.e. with F.I. smaller than 1.50) to be transported over longer distances than the flat fragments (i.e. F.I. larger than 2.00).

From the preceding, it can be concluded that fragment size plays a more important role with respect to the displacement of rock fragments than fragment shape. The large scatter, as shown in fig.3, with no distinct tendency for the small round pebbles to be transported over longer distances than the large flat cobbles, suggests that rock fragment entrainment in steep, narrow rill channels is relatively aselective. This is probably due to the fact that bedload transport in rills occurs in pulses or waves during peak flow. Hence, all rock fragments lying on the rill bed undergo a (lateral) shock independently of their size or shape. This mode of bedload transport is similar to bedload transport in desert floods, as described by SCHICK (1970) and SCHICK et al. (1985). Rill bed roughness also explains partly the aselective nature of rill flow: pools occurring in the rill beds trapped often a considerable amount of traced rock fragments.

The average distance moved in a rill during one or two flow events (tab.2) in the present field study, was about 1950 cm for a median rock fragment diameter of about 2.0 cm (fig.6). This is a movement of 975 rock diameters, a value being one order of magnitude larger as H. EINSTEIN's estimate: i.e. a single bedload movement in a river flow event is usually of the order of 100 grain diameters (LEOPOLD & EMMETT 1981).

3.3.2 Rill Site Characteristics

In order to explain the variation in observed transport distances of the traced rock fragments between the different rill sites, for each site mean displacement distances corresponding to two rock fragment diameters, i.e. 1 cm and 4 cm, were deduced from the scatter diagrams using linear regression equations. In doing so, part of the variance, due to rock fragment diameters, was removed. Next, both displacement distances were related to a set of rill site characteristics (fig.4):

$q(cm^2/s)$ peak rill flow unit discharge, estimated with the rational formula;

S mean slope gradient of the rill bed;

$w(g/s^3)$ stream power per unit boundary area, corresponding to peak rill flow discharge ($w = \rho g q S$). w is the rate at which rill flow looses energy per unit boundary area (BAGNOLD 1966);

$wz(gcm/s^3)$ product of stream power and mean specific (i.e. per weight unit) potential energy of the rock fragment at a particular rill site. $z(cm)$ equals the vertical distance between a rill site under consideration and the outlet of the experimental field plot, situated at the downslope end of the colluvial fan (figs. 1, 4). The latter is arbitrarily chosen as the reference height.

Table 5 depicts the correlation coefficients, for different types of relations, between the listed rill site characteristics and the transport distances for a 1 cm ($Y1$) and a 4 cm ($Y2$) diameter rock fragment respectively. From this table, the following general conclusions can be drawn.

- Rock fragment displacement distances always correlate better with S then with q. This corroborates findings of DE PLOEY (1983) who stated that rill initiation is controlled more closely by slope angle then by slope length and/or overland flow discharge.

- Although it has been found that mean step lengths of sand-size particles related well to the stream power of the flow (e.g. GRIGG 1970), in our study w hardly explained more of the observed variation in displacement distances of coloured rock fragments than S. Nevertheless, the use of w is preferred to S since unlike S, w allows

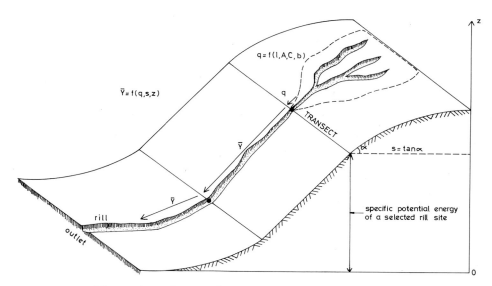

Figure 4: *Definition sketch of some rill characteristics.*
A = rill catchment area, b = bottom width of rill, C = runoff coefficient, I = rainfall intensity, q = rill flow unit discharge, s = slope gradient of rill bed, $\overline{Y}$ = mean displacement distance of traced rock fragments, z = specific potential energy of rill site.

curve type	$q - Y1$	$S - Y1$	$w - Y1$	$wz - Y1$
linear	.27	.43*	.45*	.78***
power	.55**	.68***	.68***	.49**
logarithmic	.46*	.58**	.57**	.61***
exponential	.33	.52**	.47**	.67***

curve type	$q - Y2$	$S - Y2$	$w - Y2$	$wz - Y2$
linear	.33	.44*	.51**	.78***
power	.54**	.58**	.63***	.34
logarithmic	.52**	.58***	.62***	.63***
exponential	.36	.45*	.42*	.48**

*: p = 0.10; **: p = 0.05; ***: p = 0.01
$Y1$ and $Y2$ equal displacement distances for a 1 cm and a 4 cm diameter rock fragment respectively,
q = peak unit rill flow discharge,
S = mean slope of the rill bed,
w = stream power,
z = specific potential energy of the rock fragments at a given rill site.

Table 5: *Correlation coefficients (r) for different types of relations between transport distance (Y1, Y2) of rock fragments and rill site characteristics (q, S, w, wz).*

description of the effects of different rainfall events and hence different rill flow discharges.

- Except for a power curve fit, there is a considerable increase in r when specific potential energy of the rill site is included. From the four rill site characteristics investigated, wz explained 61 percent of variation in observed Y-values. Hence, this parameter is selected as the best predictor of the observed rock fragment displacement distances (fig.5). Part of the unexplained variation in observed Y-values is probably due to a variation in rill bed roughness between the sites. For instance, in some rills, pools, occurring at regular intervals, trapped a considerable amount of traced rock fragments.

Interesting to note is that the critical slope (Scr) for rill flow transport of rock fragments, having diameters between 1 and 4 cm, varied in our study between 0.041 and 0.061. These threshold values were obtained by extrapolation of the curve, fitting the data points in an $(S) - (Y1, Y2)$ diagram for each of the 4 selected rills. This observation indicates that incision of rills in stony soils can start during moderate rainfall events on hillslopes having gradients above these Scr-values. These critical slopes are in agreement with reported Scr-values for starting rill and gully formation on colluvial fine gravels (i.e. 0.035, NEWSON 1980) as well as on loamy soils (i.e. 0.04–0.05, SAVAT & DE PLOEY 1982). In addition, these observations indicate that complete surface armoring, due to selective erosion of the fines and the concentration of rock fragments at the surface, will essentially occur on slopes less than the Scr-values mentioned. Above these critical slopes, the probability of complete surface armoring decreases.

3.4 INTENSITY OF ROCK FRAGMENT MOVEMENT BY RILL FLOW

In order to assess the highest intensity of rock fragment movement in rills during the period of observation, rock fragment discharge was calculated. Transport of rock fragment as bedload (qb) is usually expressed as an immersed weight per bed width per time unit, e.g. g/cm/s. In this study, qb was calculated with the equation:

$$qb = \sum (Wi)(Yi) \quad (6)$$

where
$Wi(g/cm^3)$ = immersed weight (= 0.62 dry weight) of available rock fragments in size fraction i,
$Yi(cm/s)$ = mean displacement distance of size fraction i per time unit.

For the calculation of maximum rock fragment discharge in a rill during the period of observation, rill site "transect green, rill 3" was selected because of the large amount of available rock fragments on the rill bed and the important displacement distances of the traced fragments at this site (fig. 3). Total weight of available bed material for transport by rill flow ($\sum Wi$) was determined after collecting the bed material, deposited during (a) previous flow event(s), from a well-defined bed surface. For the selected rill site, mean ($n = 3$) $\sum Wi$ equalled 4.4 g/cm² (or 7.1 g dry weight/cm²) and the size distribution of this bed material is shown in Fig.6. Total immersed weight of the transportable fragments larger then 2 mm, i.e. 3.7 g/cm², was divided into 8 size classes (i) and for each size class Yi was calculated using the equation (fig.3):

$$Y = -43.4D + 5011.6 (fig.3c) \quad (7)$$

where
$Y(cm)$ = mean displacement distance during rainfall event,
$D(mm)$ = fragment diameter.

Using equation (6), a qb-value of 14290 g/cm is obtained for the rainfall event causing peak runoff. This means that during the period of observation 14290 g/cm times 14 cm (= mean bottom width of a rill, table 1) = 200060 g/14 cm or ± 200 kg of rock fragments passed a mean rill cross section at transect green. If one assumes that this

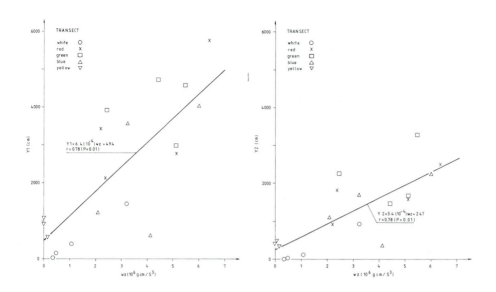

Figure 5: *Relationship between stream power times specific potential energy of a rock fragment (wz) and distance moved for a 1 cm diameter (Y1) and a 4 cm diameter (Y2) rock fragment.*

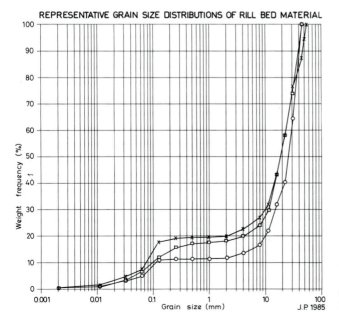

Figure 6: *Representative grain size distribution of rill bed material.*

rock fragment transport took place only during the 12 min lasting rainfall event with an intensity of 30 mm/h (tab.2), qb equals 19.8 g/cm/s or 1.98 kg per m of rill width per second. This corresponds to a dry weight discharge of 3.2 kg/m/s, a considerable figure!

Taking into account the spacing between the rills, i.e. in the mean 2.5 m at transect green, and a return period of 6 months for the rainfall event causing this important rock fragment transport, one can estimate that the maximum intensity of rock fragment movement in rills on the experimental plot equals 256 kg (dry weight) per m parallel to a contour, per year. This maximum rock fragment transport rate exceeds by two orders of magnitude maximum sediment transport rate by splash droplets on the interrills of the experimental field plot, i.e. 2.1 kg/m/year (POESEN 1985).

4 IMPLICATIONS OF RESULTS

1. This field study clearly demonstrates that under moderate rainfall conditions, rock fragments, having intermediate diameters up to 9 cm, can be transported downslope over considerable distances by rill flow. Consequently, when applying stones as a partial mulch for erosion control on rillable soils (e.g. ADAMS 1966, MEYER et al. 1972), attention should be paid to select stone sizes large enough to prevent erosion of the stone mulch itself.

2. Rock fragment content is a soil property. In addition, rock fragments themselves can be eroded by rill flow during moderate rainfall events. Hence, it would be scientifically more correct to include the effects of rock fragments on soil loss in a USLE soil erodibility factor (K) rather than in a cover and management factor (C) (WISCHMEIER & SMITH 1978, BOX & MEYER 1984).

3. From a review of the literature, THARP (1984) concluded that for natural river channels a bed is stable if the D_{85} is immobile. If we apply this principle to the rill channels at the experimental plot which have a D_{85} of 4 cm (fig.6), we can conclude from tab.3 that most of the rill beds were unstable during the recorded moderate rainfall event. Hence, it can be stated that rill channel armoring will be overcome several times a year on the experimental plot. The formation of an armor layer in rills formed in a highly gravelled soil, as described by FOSTER (1982, 338), may therefore only be limited to low magnitude rainfall events.

4. Our field results have also some implications for taphonomy. On the upper part of the experimental plot two flint artefacts, having intermediate diameters of 1.05 and 1.5 cm respectively and probably dating from the Mesolithic (VERMEERSCH, pers. comm.), were found: one on an interrill, the other in a rill. VERMEERSCH (in press) states that in the Belgian loam region, mesolithic sites are virtually unknown. Furthermore, this author assumes that if mesolithic sites were located on loam slopes, they now have disappeared due to soil erosion. Our findings give more insight into the processes which are responsible for the evacuation of artefacts from loam covered slopes: i.e. rill flow can easily transport flint artefacts with intermediate diameters up to

9 cm (tab.3) during moderate rainfall events, while the competence of inter-rill flow during such events is almost one order of magnitude smaller. Since rilling is very likely to occur on bare loam covered slopes with slope gradients larger than 0.04–0.05 (SAVAT & DE PLOEY 1982), and since forest clearance and cultivation of the soils in the Belgian loam belt started since at least medieval times, the probability for downslope transport of artefacts by rill flow in the loam region is very high.

5. The presence of rock fragments in colluvial deposits, as has been described by several authors (IMESON 1978, KWAAD & MUCHER 1979, PAULISSEN et al. 1981, V.D. BRINK & JUNGERIUS 1983, BORK 1985) is an important pointer to the type of processes acting on the upland areas during the period of colluviation. In fact, maximum size of rock fragments found in colluvial deposits reflects the competence of the overland flow which transported them downslope, provided that overland flow is the responsible process and provided that a broad range of fragment sizes was available on upslope sections at the time of upland erosion. This fragment size can than be used to reconstruct paleohydrological conditions. For instance, for a given upland slope gradient (S) the critical discharge (q_c) can be calculated using equation (5). Using equation (3) together with the calculated q_c-value and the competent fragment diameter (D), deduced from the rock fragment size distribution in the colluvial deposit, an estimate of rill or gully flow unit discharge (q) can be made. Knowing slope length and assuming a value for the runoff coefficient and a value for the mean width of the rill or gully catchment, an estimate of peak rainfall intensity can then be made. Equation (5) must be seen as a first step towards a relation between S and q_c. More field and laboratory data are needed to confirm and extend this relation.

6. From the field observations it can be safely said that rill and/or (ephemeral) gully flows are the main processes responsible for the evacuation of rock fragments from upland areas in the Belgian loam region. Since more and more gravelly deposits are outcropping in this area due to upland erosion or due to ploughing on shallow loam soils, it can be expected that more and more rock fragments will be evacuated and redistributed downslope by rill and/or (ephemeral) gully flow. Hence, the area with top soil containing rock fragments is increasing. Part of the eroded rock fragments will be transported by these overland flow processes to the river system. Hence, it can be expected that, as long as rilling and/or gullying occurs in the study area, this new type of sediment can change stream response. For instance, it has been shown that an increase of gravel input to the stream system provokes channel widening (e.g. GRISSINGER & MURPHEY 1982, HARVEY 1985).

5 CONCLUSIONS

The main conclusions from this field study can be summarized as follows:

1. Rill flow, generated during moderate rainfall events, can be identified as the most important process leading to the downslope movement of rock fragments on upland areas in the Belgian

loam region. The competence of rill flow during such events exceeds almost by a factor 10 the competence of interrill flow. Obviously, during extreme rainfall events rill flow will be even more effective in moving rock fragments.

2. Incipient motion conditions for single or clustered rock fragments lying on a rill bed, coincide with a SHIELDS criterion value (Θ_c) of 0.015. This Θ_c-value is smaller than the general accepted Θ_c-value for conventional open channel flows with gentle slopes, large ratios of flow depth to sediment size and fine bed materials (i.e. $\Theta_c = 0.05$).

3. With respect to the influence of rock fragment properties on the distance rock fragments moved by rill flow, it can be stated that fragment size plays a more important role than fragment shape. Nevertheless, for a given fragment diameter a large variation in transport distance is observed, suggesting that rock fragment transport by rill flow occurs in pulses or waves. In addition, rill bed roughness also accounts for the observed variation. Considering several rill site characteristics, it was observed that rock fragment displacement distances always correlate better with rill bed slope than with peak rill flow unit discharge and, hence, also rill catchment area. Critical slope gradient for rill flow transport of 1 and 4 cm diameter rock fragments varied between 0.041 and 0.061. The product of stream power and mean specific potential energy of the rock fragment at a particular rill site explained 61 per cent of the variation in mean transport distance for a 1 and 4 cm diameter rock fragment between the rill sites.

4. During a moderate rainfall event, maximum transport of rock fragments as bedload (q_b) can amount to 200 kg (submerged weight) per rill cross section on the field plot. Taking into account the mean distance between the rills as well as the return period of the rainfall causing peak runoff, this figure corresponds to a maximum intensity of rock fragment transport of 256 kg (dry weight) per m contour length and per year. This transport rate exceeds by two orders of magnitude maximum transport rate by raindrop splash on the field plot.

ACKNOWLEDGEMENT

I am indebted to Prof. J. de Ploey who provided for the field plot in Huldenberg and to Prof. P.M. Vermeersch for a discussion on the implications of the results for archaeology. Dr. G. Govers, Drs. G. Rauws and Mr. L. Cleeren are thanked for their assistance in the field. I also wish to thank Drs. G. Wyseure who put the rainfall records at my disposal. Mr. R. Geerarts is thanked for drawing the illustrations and Ms. A. Van Elsen for typing the manuscript.

REFERENCES

[ADAMS 1966] ADAMS, J.: Influences of mulches on runoff, erosion and soil moisture depletion. Soil Science Society America Proceedings **30**, 1966, 110–114.

[ANDREWS 1983] ANDREWS, E.D.: Entrainment of gravel from naturally sorted riverbed material. Geological Society of America Bulletin **94**, 1983, 1225-1231.

[BAGNOLD 1966] BAGNOLD, R.A.: An approach to the sediment transport problem from general physics. Geological Survey Professional Paper **422-I**, 1966, 37 p.

[BATHURST et al. 1982] BATHURST, J.C., GRAF, W.H., CAO, H.H.: Initiation of sediment transport in steep channels with coarse bed material. In: MUTLU SUMER, B. & MULLER, A. (eds.), Mechanics of Sediment Transport. Balkema, Rotterdam, 1982, 207–213.

[BETTES 1984] BETTES, R.: Initiation of sediment transport in gravel streams. Proc. Instn. Civ. Engrs. Part 2, **77**, 1984, 79–88.

[BORK 1985] BORK, H.R.: Mittelalterliche und neuzeitliche lineare Bodenerosion in Südniedersachsen. Hercynia N.F. **22**, 1985, 259–279.

[BOX & MEYER 1984] BOX, I.E. & MEYER, L.D.: Adjustment of the Universal Soil Loss Equation for Cropland. Soils Containing Coarse Fragments. SSSA Special Publication **13**, 1984, 83–90.

[DE PLOEY 1983] DE PLOEY, J.: Runoff and rill generation on sandy and loamy topsoils. Zeitschrift für Geomorphologie Suppl.Bd. **46**, 1983, 15–23.

[DE PLOEY & MOEYERSONS 1975] DE PLOEY, J. & MOEYERSONS, J.: Runoff creep of coarse debris: experimental data and some field observations. CATENA **2**, 1975, 275–288.

[FENTON & ABBOTT 1977] FENTON, J.D. & ABBOTT, J.E.: Initial movement of grains on a stream bed: the effect of relative protrusion. Proc. R. Soc. London **352**, 1977, 523–537.

[FOSTER 1982] FOSTER, R.G.R.: Modeling the erosion process. In: HAAN, C.T. (ed.), Hydrologic Modeling of Small Watersheds. ASAE Monograph **5**, 1982, 297–380.

[GOVERS & POESEN 1986] GOVERS, G. & POESEN, J.: A field-scale study of surface sealing and compaction on loam and sandy loam soils. Part I. Spatial variability of soil surface sealing and crusting. In: CALLEBAUT, D., GABRIELS, D. & DE BOODT, M. (eds.), Assessment of Soil Surface Sealing and Crusting, 1986, 171–182.

[GRIGG 1970] GRIGG, N.S.: Motion of single particles in alluvial channels. Journal of the Hydraulics Division HY **12**, 1970, 2501–2518.

[GRISSINGER & MURPHEY 1982] GRISSINGER, E.H. & MURPHEY, J.B.: Present "Problem" of stream channel instability in the Bluff Area of Northern Mississippi. Journal of the Mississippi Academy of Sciences **27**, 1982, 117–128.

[HARVEY 1985] HARVEY, A.M.: Sediment supply to upland streams from eroding gullies. Abstract of paper presented at the Int. Workshop on Problems of Sediment Transport in Gravel-bed Rivers, Pingree Park, Colorado, 12–15 August 1985.

[IMESON 1978] IMESON, A.C.: Slope deposits and sediment supply in a New England drainage basin (Australia). CATENA **5**, 1978, 109–130.

[KWAAD & MUCHER 1979] KWAAD, F.J.P.M. & MUCHER, H.J.: The formation and evolution of colluvium on arable land in northern Luxembourg. Geoderma **22**, 1979, 173–192.

[LAURANT 1976] LAURANT, A.: Nouvelles recherches sur les intensités maximums de précipitations à Uccle. Courbes d'intensité-durée-fréquence. Annales des Travaux Publics de Belgique **4**, 1976, 320–328.

[LEOPOLD & EMMETT 1981] LEOPOLD, L.B. & EMMETT, W.W.: Some observations on the movement of cobbles on a streambed. Proceedings of the IAHS Symposium, Erosion and Sediment Transport measurement, Firenze, Italy, 1981, 49–59.

[MEYER et al. 1972] MEYER, L., JOHNSON, C, & FOSTER, G.: Stone and woodchip mulches for erosion control on construction sites. Journal of Soil and Water Conservation **27**, 1972, 264–269.

[MILLER & GUTHRIE 1984] MILLER F.T. & GUTHRIE, R.L.: Classification and distribution of soils containing rock fragments in the United States. SSSA Special Publication **13**, 1984, 1–6.

[MOEYERSONS & DE PLOEY 1976] MOEYERSONS, J. & DE PLOEY, J.: Quantitative data on splash erosion, simulated on unvegetated slopes. Zeitschrift für Geomorphologie Suppl.Bd. **25**, 1976, 120–131.

[NEWSON 1980] NEWSON, M.D.: The erosion of drainage ditches and its effect on bedload yield in mid-Wales: reconnaissance case studies. Earth Surface Processes **5**, 1980, 275–290.

[NICHOLS et al. 1984] NICHOLS, J.D., BROWN, P.L. & GRANT, J.: Erosion and Productivity of Soils containing Rock Fragments. SSSA Special Publication **13**, 1984, vii.

[NIELSEN 1984] NIELSEN, D.R.: Foreword in: Erosion and Productivity of Soils containing Rock Fragments. SSSA Special Publication **13**, 1984, v.

[PAULISSEN et al. 1981] PAULISSEN, E., GULLENTOPS, F., VERMEERSCH, P.M., GEURTS, M.A., GILOT, E., VAN NEER, W., VAN VOOREN, E. & WAGEMANS, E.: Evolution holocène d'un flanc de vallée sur substrat perméable (Hesbaye sèche, Belgique). Mémoires de l'Institut Géologique de l'Université de Louvain **21**, 1981, 23–73.

[POESEN 1985] POESEN, J.: An improved splash transport model. Zeitschrift für Geomorphologie **29**, 1985, 193–211.

[POESEN & GOVERS 1986] POESEN, J. & GOVERS, G.: A field-scale study of surface sealing and compaction on loam and sandy loam soils. Part II. Impact of soil surface sealing and compaction on water erosion processes. In: CALLEBAUT, D., GABRIELS, D. & DE BOODT, M. (Eds.), Assessment of Soil Surface Sealing and Crusting. 1986, 183–193.

[SAVAT & DE PLOEY 1982] SAVAT, J. & DE PLOEY, J.: Sheetwash and rill development by surface flow. In: BRYAN, R. & YAIR, A. (eds.), Badland Geomorphology and Piping. Geo Books, Norwich, 1982, 113–126.

[SCHICK 1970] SCHICK, A.P.: Desert floods — interim results of observations in the Nahal Yael research watershed, 1965–1970. International Association of Scientific Hydrology, Publication **96**, 1970, 478–493.

[SCHICK et al. 1985] SCHICK, A.P., LEKACH, J.& HASSAN, M.A.: Bedload transport in desert floods: observations in the Negev. Abstract of paper presented at the Int. Workshop on Problems of Sediment Transport in Gravel-bed Rivers, Pingree Park, Colorado, 12–15 August 1985.

[SIMANTON et al. 1984] SIMANTON, J.R., RAWITZ, E. & SHIRLEY, E.D.: Effects of rock fragments on erosion of semiarid rangeland soils. SSSA Special Publication **13**, 1984, 65–72.

[THARP 1984] THARP, T.M.: Sediment characteristics and stream competence in ephemeral intermittent streams, Fairborn, Ohio. CATENA Suppl.Bd. **5**, 1984, 121–136.

[V.D.BRINK & JUNGERIUS 1983] VAN DEN BRINK, J.W. & JUNGERIUS, P.D.: The deposition of stony colluvium on clay soil as a cause of gully formation in the Rif Mountains, Morocco. Earth Surface Processes and Landforms, **8**, 1983, 281–285.

[VANONI 1977] VANONI, V.A.: Sedimentation Engineering. ASCE Manuals and Reports on Engineering Practice-No. **54**, 1977, 745 p.

[VERMEERSCH] VERMEERSCH, P.M.: Ten years research on the mesolithic of the Belgian lowland: results and prospects. In press: Proc. III Int. Symp. on the Mesolithic in Europe, Edinburgh (1985).

[WISCHMEIER & SMITH 1978]
WISCHMEIER, W.H. & SMITH, D.D.: Predicting rainfall erosion losses, U.S.D.A., Agriculture Handbook No. **537**, 1978, 58 p.

Address of author:
J. Poesen
Research Associate, National Fund for Scientific Research,
Laboratory of Experimental Geomorphology, K.U. Leuven
Redingenstraat 16 bis,
B-3000 Leuven.

RILL DEVELOPMENT IN A WET SAVANNAH ENVIRONMENT

O. **Planchon**, E. **Fritsch** & C. **Valentin**, Abidjan

SUMMARY

A 136 ha-watershed, representative of the wet savannah environment, was selected in northern Ivory Coast to analyse rill development in relation to environmental circumstances. Several detailed maps were produced by an interdisciplinary team indicating the distribution of soils, vegetation cover, sealed surface areas, land use together with geomorphology and rill patterns. In addition, 81 linear incisions, including pre-rills, rills, gullies and deep gullies, were surveyed thus enabling a quantitative analysis, based upon the length and the volume of incision. Rill depth is mainly controlled by slope inclination and the density of incisions by slope length and permeability. Three distinctive rill systems could be distinguished within the catchment:

- On upper slopes pre-rills, rills and gullies depend upon surface flow which in turn is governed by soil surface features.

- Midslope, gullies and deep gullies are mainly induced by the steeper slope. Pedological analysis shows that lowering of land surface has been partly originated by pedological processes of lateral leaching which, besides, have induced some forms of piping erosion.

- The downstream system consists of rills developed on clay deposits and joins the brook, but is not connected to the other two systems. except for heavy rainstorms, this downslope system starts operating in the late rainy season once the foothill watertable becomes sufficiently shallow to permit hydraulic continuity of the overall rill pattern.

RESUME

Un bassin versant de 136 ha, représentatif de la savane humide du nord de la Côte d'Ivoire, a fait l'objet d'une étude sur les facteurs d'érosion linéaire. A la suite d'un travail interdisciplinaire, plusieurs cartes détaillées ont été dressées concernant les sols, la végétation, les organisations pelliculaires superficielles, l'utilisation des terres ainsi que le relief et le réseau de drainage. De plus, les profils de 81 incisions linéaires ont été relevés, ce qui a permis une étude quantitative fondée sur la longueur et le volume d'incision; plusieurs formes d'érosion linéaire ont été distinguées: proto-griffes, griffes, ravinaux et ravines. La profondeur des incisions dépend surtout de la pente alors que leur densité est davantage reliée à la longueur de pente et à la perméabilité. Trois systèmes d'érosion linéaire ont pu être différenciés:

- Celui de haut de versant, constitué de proto-griffes, de griffes et de ravinaux, dépend du ruissellement superficiel et des conditions de surface.

- A la mi-versant, la formation de ravinaux et de ravines est induite par une pente plus marquée. L'analyse pédologique montre que l'abaissement topographique est à imputer, pour une bonne part, à des processus pédologiques de soutirage qui se manifestent notamment par des formes d'érosion en tunnel.

- En bas de versant, des griffes entaillent des dépôts colluvio-alluviaux argileux. Ce troisième système rejoint le marigot mais n'est pas relié morphologiquement aux deux systèmes précédents. Exception faite des forts orages, ces griffes ne deviennent fonctionnelles qu'en fin de saison des pluies, lorsque la remontée de la nappe de bas de versant assure la continuité hydrologique de l'ensemble du système de drainage.

1 INTRODUCTION

Rill erosion may be an extremely severe form of erosion in regions exposed to heavy rainfall such as the wet tropics. Among the variables controlling rill development, the overriding importance of slope-induced hydraulic conditions has been recognized in many studies (MOSLEY 1974, KALMAN 1976, SAVAT & DE PLOEY 1982, GOVERS 1985). The important influence of soil and bed rock characteristics have also been emphasized (SAVIGEAR 1960, ROLOF et al. 1981). Because of the complexity of the interacting factors involved, full understanding of rill development still remains a difficult task. It is necessary first to clearly define the terms which will be used. As widely accepted, a distinction will be made between interrill erosion resulting mainly from detachment by raindrop impact, and rill erosion resulting from detachment by concentrated flow. Furthermore four types of incision were distinguished:

1. Pre-rills: due to the irregular microtopography, surface runoff does not flow parallel to the slope but meanders and anastomoses. Where discharge increases, flow tends to concentrate in straighter channels among the tufts of grass. When linear flow was not associated with conspicuous traces of incision but only affected the few top millimeters of soil surface, the feature was termed "pre-rill". It was considered important to identify this minor form of linear erosion since it was assumed that it represented the initial stage in rill development.

2. Rills: shallow and narrow channels that only affect topsoil layers (depth < 0.15 m).

3. Gullies: water courses cut into subsoil. These are formed with sloping sides and V-shaped bottoms (0.15 m < depth < 1 m).

4. Deep gullies: differ from ordinary gullies not only by their depth (1 m) but also by their U-shaped or inverted T-shaped bottoms. Several headcuts usually occur along their profile. Moreover, the low slope

inclination of the gully floor does not reflect the landform of the gully catchment.

This paper describes relationships between the evolution of linear incisions and the controlling factors.

2 GENERAL DESCRIPTION OF THE BOORO-BOROTOU WATERSHED

When selecting the catchment, it was prerequisite that its environmental characters was representative of the north west of Ivory Coast, in an area where soils had recently been mapped (1/200,000 scale). The Booro-Borotou, 136 ha-watershed meets this requirement well. It is located about 25 km north of Touba. Mean annual rainfall recorded over a period of 33 years, is 1359 mm (standard deviation: 220 mm), of which 69% falls during the rainy season (BROU 1986).

Except during dry years, base flow continues throughout the dry season. The part of the total flow which is due to the drainage of the water table largely exceeds runoff flow which only represents 15.3% in 1984, and 17.3% in 1985 of the total (CHEVALLIER, pers. comm.), which is common in this climatic zone (DUBREUIL 1985).

The bedrock consists of undulated injection gneiss. A fracture network characterized by joints and bedding-planes is perpendicular to the stream. Main geomorphic features are presented in fig.1. Due to a capping iron pan, the four remnants of plateaus are scrapped with concave slopes up to 22°. The upper slope segment is generally rectilinear with a gentle slope (1°). A slope break occurs at midslope where indurated layers, together with an iron pan, outcrop locally. Downslope segments are convex-concave with inclination of 3°. In foothill areas, concave depressions, perpendicular to the stream, have developed mainly on the right bank, which join the stream with very low slope angle. The valley floor is generally concave with slope gradient of 3°.

A detailed pedological study was carried out. The three-dimensional structure of the soil mantle was investigated, describing and sampling nine toposequences. Complementary observations were required to unravel the great complexity of the lateral variations between the toposequences. Three main pedological domains were differentiated (tab.1):

1. in upper hillslopes, a ferrallitic domain comprises four remnants of cuirassed plateaus and their surroundings.

2. a ferruginous domain which covers the most part of the watershed. From a classifying viewpoint, these soils may be considered as ferrallitic soils grading into ferruginous soils, as they will be termed hereafter. They are affected by two main processes:

   • impoverishment in iron and clay that originates from the topsoil. Downstream this is emphasized in association with pronounced degradation of the physical properties.

   • accumulation of iron in the deeper layers which result in the development of indurated layers which outcrop locally midslope.

3. Downslope, the hydromorphic domain is related to the permanent

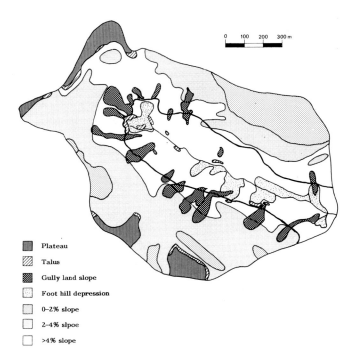

Figure 1: *Map of the main landform features.*

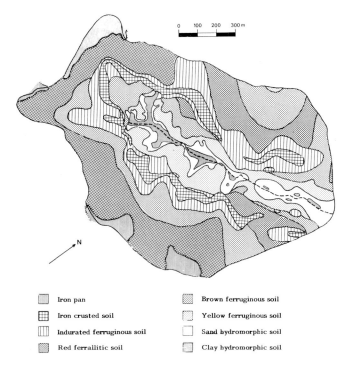

Figure 2: *Soil map.*

Rill Development, Wet Savannah

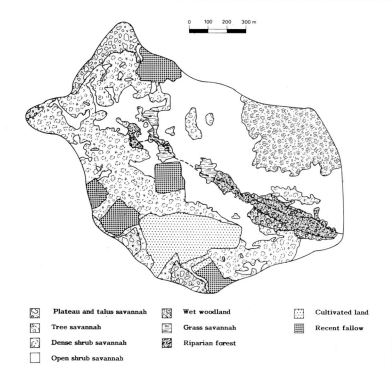

	Plateau and talus savannah		Wet woodland		Cultivated land
	Tree savannah		Grass savannah		Recent fallow
	Dense shrub savannah		Riparian forest		
	Open shrub savannah				

Figure 3: *Vegetation, land use map.*

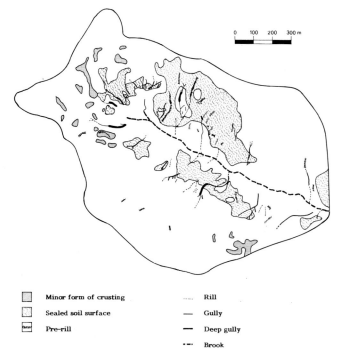

	Minor form of crusting	.....	Rill
	Sealed soil surface	—	Gully
	Pre-rill	—	Deep gully
		-·-	Brook

Figure 4: *Surface features and rill pattern map.*

Soil	Texture of the topsoil	Clay content of the topsoil %	Texture of the subsoil	Maximum distance to the water divide (m)	Mean percent slope
Ferrallitic					
Iron pan	Sandy Clay Loam	25	Iron pan	200	2
Red Ferrallitic	Sandy Clay Loam	38	Clay	400	6
Ferruginous					
Brown ferruginous	Sandy Loam	14	Sandy Clay Loam	550	7
Indurated ferruginous	Sandy Loam	17	Indurated	490	6
Iron crusted	Sandy Clay Loam	25	Iron crust	460	8
Yellow ferruginous	Sand	06	Loamy Sand	600	7
Hydromorphic					
Sand hydromorphic	Sandy Loam	11	Sand	580	6
Clay hydromorphic	Silty Clay Loam	26	Clay	580	5

Table 1: *Main characteristics of the soil types.*

Land unit	Range of cover over year (%)	Maximum distance to the water divide (m)	Mean percent slope
Plateau and talus savannah	15–90	200	12
Tree savannah	40–90	360	1
Dense shrub savannah	30–90	450	6
Open shrub savannah	10–90	550	6
Savannah with sealed soil surface	05–80	410	7
Wet woodland	50–100	540	7
Grass savannah	50–100	600	6
Riparian forest	90–100	570	9
Cultivated land	<5–90	400	1
Recent fallow	05–80	550	5

Table 2: *Main characteristics of the land units.*

water table which drains towards the stream exporting first iron, then clay. As a result a thick gray sand layer develops and may locally extend upward to the midslope iron pan. In such a location, this layer cannot result from colluvial processes. Downstream from the depression and in the valley bottom, fine black colluvial-alluvial deposits overlie this sand layer.

The distribution of soils within the catchment (fig.2) results from material transfers. Accumulation of iron which caps midslope areas occurs precisely where regolithic layers are shallowest. Lateral leaching of iron and clay originates from soil surface upslope and from deep layers downslope. Such processes should be responsible for overall lowering of the land surface, downstream of the midslope iron pan. Evidence is based on the occurrence of **in situ** relict regolith within the leached sandy layers and on structural relationships between these layers and the watertable. Such volume decrease due to internal lateral leaching has been documented in Burkina Faso by BOULET et al. (1977).

A specific study of the surface features was undertaken including vegetation, land use and soil surface differentiations such as micro-topography and surface seal. Six main vegetation map units were identified (fig.3).

Owing to the vegetation cover and faunal activity, soil surface sealing is mainly a seasonal phenomenon in most of the watershed. However, the most severely affected areas were mapped (fig.4). Primary forms of surface degradation occurs upslope within finger-shaped glades which penetrate into the dense shrub savannah and the surface of which is slightly sealed. These glades are continuous with more severely crusted surfaces which occur within the open shrub savannah. These areas correspond to a further stage of surface degradation and are characterized by grass tussocks with a soil level above the surrounding surface (the difference in elevation varies from 3 to 15 cm). The bare patches (0.2 to 1.0 cm diameter) are severely crusted and infiltrability is limited to 10 mm/hr under simulated rainfall (phot.1). The ultimate stage of surface change occurs downstream, where surface roughness gradually increases so that microtopography forms step-like features.

Only a small part of the catchment is cultivated, but because of annual anthropogenic bush fires, vegetation recovers slowly after cultivation. Features associated with short fallow (less than 7 years) are easily recognizable in the fields (fig.3). Finally, 10 land units have been differentiated (tab.2).

3 THE LINEAR EROSION PATTERNS

3.1 THE CHANNEL NETWORK

Linear erosion is a frequent phenomenon within the watershed (fig.4). The channel network is not digitate but axial; usually, individual incisions have no tributary, or if any, very short one. The rill network is discontinuous. Upslope and midslope rills are generally not morphologically connected to the downslope rills which occur on the clay hydromorphic soil and which join the stream.

Most frequently the incision depth, hence the rill type, varies along the slope: moving from upslope towards the valley bottom, a pre-rill which originated within a sealed soil surface may turn into rill, gully or deep gully before vanishing in the sand hydromorphic soil and reappearing in the form of a rill in the clay hydromorphic soil. This complexity will be illustrated through an example (fig.5).

The selected incision drains a 11.1 ha catchment. The distance from the pre-rill head to the water divide is 320 m. The variations along the slope of the longitudinal and perpendicular cross sections are presented in fig.6. Several portions can be distinguished: upslope, a 30 m long pre-rill developed in a 2 years old cassava field. When cutting into the ridges of a cotton field, it changes into a 20 cm-deep and 2 m-broad rill. Downstream, between 90 and 160 m from the pre-rill head, the uncultivated soil gradually becomes more indurated and is dissected by a 50 cm-gully. At 160 m, a first headcut occurs between this gully and a 1.30 m-deep gully. It corresponds to the limit between the midslope iron pan and the less resistant brown ferruginous soil. A second headcut is located 20 m

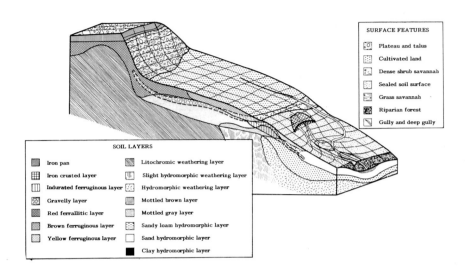

Figure 5: *Block diagram illustrating the rill example.*

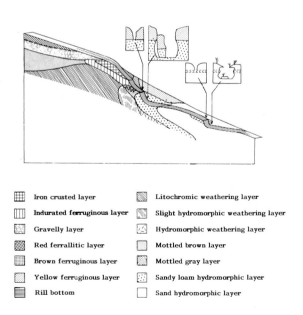

Figure 6: *Longitudinal section of the rill example, with four cross sections.*

Photo 1: *Sealed soil surface associated to typical microtopography.*

Photo 2: *The deepening and the broadening of the example rill when reaching the depression.*

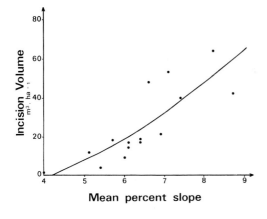

Figure 7: *The impact of the mean percent slope on the incision volume.*

downstream; the incision deepens to 3 m where the lateral leaching of iron and clay make the soil materials even less resistant. The gully banks remain vertical wherever they cut into the sandy gray layer but they are undermined where incision also affects the hydromorphic materials. Such mass wasting may be ascribed to the sapping of the watertable. As shown in fig.5, the gully direction clearly deviates towards the centre of a depression precisely where it enters the zone affected by the water-table, namely where soils are subject to the ultimate lateral leaching. The inclination of the gully bottom, 1.8°, is lower than the steepness of the ground surface, the gully depth gradually decreases from 3 to 0.8 m, between 180 and 260 m. When joining to the depression, the gully deepens to 1.4 m and broadens from 0.8 to 2 m (phot.2). Its sides are severely caved in. Gradually, the gully depth decreases so that at 300 m incision is no longer perceptible (phot.3). At 310 m, entering in the grass savannah where the water-table ocasionally reaches the surface, an anastomosed rill system is observed which is channelled and deepens to 0.60 cm at 340 m where the riparian forest occurs before joining the stream at 370 m.

Photo 3: *Sand deposits in the low section of the example rill before turning into deeper rill.*

4 QUANTITATIVE STUDY

4.1 METHODS

Quantitative variables include those describing environmental conditions (tab.1 and 2) and those describing gully characteristics (tab.3).

In each soil and land unit, the mean slope inclination was assessed from the 1/2,500 topographic map using the grid point sampling technique described further. The variations of soil cover (tab.2) along the year was estimated using detailed vertical photographs taken at 5 m above ground level and pint points method in the field. In each unit, the maximal distance from the water divide was measured on the 1/2,500 maps. Moreover, in order to use the model of BOON & SAVAT (1981), mean grain size was determined for each unit.

In the field, 81 linear erosion features were topographically surveyed. In each,depth, groundlevel and bottom widths were measured enabling the computation of the cross section, considered as trapezoid. Since these measurements were performed every 10 m along the incisions, their volume and length could be rather accurately evaluated. In ad-

		Pre-rill	Rill	Gully	Deep gully
Depth (m)	Mean	0	0.1	0.6	1.5
	Standard deviation	0	0.1	0.8	0.6
Cross section (m^2)	Mean	0	0.1	0.8	3.1
	Standard deviation	0	0.2	1.0	2.2
Length (m)	Total	1250	2780	1190	500
Volume (m^3)	Total	0	274	915	1470
Slope percent parallel to the incision	Mean	5	6	8	12
	First quartile	3	4	5	5
	Standard deviation	3	3	6	11
Distance to the water divide (m)	Mean	304	358	396	385
	First quartile	220	270	280	300
	Standard deviation	111	116	147	105

Table 3: *Main characteristics of the incisions.*

	Iron crusted soil	Yellow ferruginous soil	Clay Hydromorphic soil	Brown ferruginous soil	Sand Hydromorphic soil	Indurated ferruginous soil	Red ferrallitic soil	Mean
Mean cross section (m^2)								
Rill	0.11	0.07	0.07	0.14	0.05	0.14	0.13	0.10
Gully	0.77	1.06	0.44	1.09	0.68	0.61	0.75	0.77
Deep gully	4.62	3.80	-	2.27	1.40	3.60	3.80	2.94
Volume (m^3/Ha)								
Rill	5	3	12	2	2	4	1	2
Gully	24	18	-	7	4	7	3	7
Deep gully	35	19	-	11	13	7	10	11
Total	64	40	12	20	18	18	14	20
Length (m/Ha)								
Pre-rill	27	14	9	11	-	11	7	9
Rill	40	39	163	15	29	28	8	20
Gully	31	17	-	6	6	11	4	9
Deep gully	8	5	-	5	9	2	3	4
Total	106	75	172	37	44	52	22	42
Slope percent parallel to incision								
Pre-rill	6	6	5	6	-	5	5	5
Rill	5	7	5	6	6	6	6	6
Gully	9	9	3	7	3	6	8	8
Deep gully	28	11	-	10	7	10	6	11

Table 4: *The influence of soil type on the rill characteristics.*

dition, slope inclination parallel to the rill was measured every 10 m. To compare the influence of the different environmental factors, length and volume were expressed per hectare. These variables were combined with the type of erosion feature: prerill, rill, gully or deep gully (tab.3).

The relationships between linear erosion forms and environmental parameters were determined using the field data together with data collected by simultaneous sampling of the topographic, soil and surface features maps. Data from the maps were collected at 2,200 sample points (16 points/ha) using the grid point sampling technique. The frequency of each relationship was calculated as a percentage of the whole sample (tab. 4 and 5).

	Sealed soil surface	Wet woodland	Riparian forest	Dense shrub savannah	Open shrub savannah	Grass savannah	Recent fallow	Cultivated land	Mean
Mean cross section (m²)									
Rill	0.13	0.10	0.04	0.14	0.10	0.07	0.10	0.08	0.10
Gully	0.85	0.82	0.60	0.73	1.20	0.67	0.77	-	0.77
Deep gully	3.70	1.56	3.55	3.63	1.18	-	-	-	2.94
Volume (m³/Ha)									
Rill	5	3	1	2	2	9	0	1	2
Gully	34	18	14	7	10	3	4	0	7
Deep gully	14	27	27	10	5	0	0	0	11
Total	53	48	42	19	17	9	4	1	20
Length (m/Ha)									
Pre-rill	31	6	0	5	11	0	4	6	9
Rill	40	37	35	14	25	121	4	14	20
Gully	17	22	24	9	8	4	5	0	9
Deep gully	9	17	8	3	4	0	0	0	4
Total	97	81	67	31	48	125	13	20	42
Slope percent parallel to incision									
Pre-rill	6	5	-	5	6	-	3	4	5
Rill	7	6	6	5	7	6	6	4	6
Gully	7	7	8	9	7	5	8	-	8
Deep gully	13	8	25	10	5	-	-	-	11

Table 5: *The influence of land unit on the rill characteristics.*

5 RESULTS AND DISCUSSION

5.1 SLOPE-INDUCED HYDRAULIC CONTROL

For units affected by rill erosion, a procedure of step-wise multiple regression was adopted whereby the least significant contributory variables (based on t-value) are progressively dropped.

The most significant individual variable on rill volume is slope steepness: $n=15$, $r=0.82$. The other variables are randomly correlated with rill volume and were eliminated. Mean slope steepness of each unit appears as a major controlling factor since 67% of the incision volume variations can be accounted for regardless of soil type and land unit. The regression curve (fig.5) indicates that no rill develops on units where mean slope inclination is below a threshold value of 2° (i.e. 4%) which agrees closely with the observations of SAVAT & DE PLOEY (1982).

As shown in tab.3, the slope steepness parallel to incisions influences the depth of incision. Below 4° (7%), incisions remain shallow (pre-rills and rills), between 4° and 6° (7% and 10%), gullies occur whereas deep gullies develop on slopes steeper than 6°.

The variables tested were not well correlated individually with rill density, expressed as length of incision per hectare. However, a two-variable equation containing maximal length to the water divide and clay content of the top soil gives a significative correlation coefficient of 0.66 ($n=15$).

These results may be discussed in terms of hydraulic parameters. As shown by several authors (SAVAT 1979, PLANCHON 1985), rills can only start when all particle size fractions can be equally removed by a critical flow verlocity which depends on slope inclination and unit discharge. Our results suggest that incision depth is mainly controlled by slope gradient whereas rill density is influenced by unit discharge, assuming that unit discharge depends upon slope length and clay content, as assessed by BOON & SAVAT (1981).

The nomographs of BOON & SAVAT (1981) permit prediction of the initiation of rill incision. They require data

Photo 4: *Sediment deposits from sheet runoff erosion at the bottom of a cultivated field.*

on slope length, slope steepness, median grain size and clay content of the top soil. In our case, results match the predictions for 10 units out of 18. Discrepancies may be ascribed to two main sources of error:

1. incorrect estimation of permeability: rill should not develop either in the sealed surface unit in the fringe forest since both estimated permeabilities exceed 250 mm/hr, but both units are among the most affected by linear erosion. This is probably enhanced by runoff factors which are more favourable than can be predicted by textural conditions. In the case of the sealed surface unit, infiltrability assessed under rainfall simulation is only 10 mm/hr, whereas drainage conditions in the fringe forest are seriously impaired by the permanent water-table. Conversely, despite predictions, no rilling is observed on the iron pan, plateau and talus savannah units. Infiltrability on the talus savannah unit under simulated rainfall is 15 mm/hr (CHEVALLIER & SAKLY 1985) exceeding the predicted permeability (6 mm/h) because clay is mainly found in the form of water-stable aggregates.

2. heterogeneity of units along the hillslope. According to their intrinsic characteristics, the brown and yellow ferruginous soils together with the open shrub savannah and the riparian forest should not be exposed to linear erosion hazards. The fact that they are actually eroded may be therefore attribute to external factors: they receive surface flow from runoff contributing areas located upstream.

5.2 SOIL COVER

No clear influence of soil cover on linear erosion has been established. Incomplete soil cover, as for the sealed surface, enhances sealing and hence discharge, favouring incipient linear erosion. Yet, once rill flow has initiated from upstream, the canopy is largely unsuccessful in eliminating it. Even the most complete soil cover comprising high trees (30 m high) and dense understorey are ineffective since the riparian forest and the wet woodland are among the most severely affected units (tab.5).

5.3 LAND USE

The question arises whether cultivation accelerates rill erosion. On the one hand, cultivated and fallow lands are not severely affected by linear erosion (tab.5). This result may be attributed to the land use system which is rather conservative. The red ferrallitic soils which are mainly selected by the farmers for clearance, present low linear erosion hazard. Owing to their intrinsic properties, they fall on the "no risk class" of the BOON & SAVAT (1981) nomographs. Moreover, the duration of cultivation is limited to 5 to 7 years, interrupting irreversible processes of soil surface degradation. On the other hand, fields are affected by runoff and sheet erosion as indicated by sand deposits (phot.4). Besides, it has been shown that tillage, like ridging, may accelerate linear erosion. Even though land degradation remains limited within the cropped lands, cultivation can be a contributory factor on other soils as suggested by the important gully located downstream from the cotton fields.

6 RILL DEVELOPMENT SYSTEMS

Quantitative data together with structural observations enabled us to identify three rill systems within the watershed.

6.1 UPSLOPE SYSTEM

Under natural conditions a complex runoff system develops upslope due to gradual degradation of land surface conditions along the hillslope. Change in infiltration conditions starts in the elongated glades within the dense shrub savannah (fig.4). Although the soil surface is only slightly sealed, observations on runoff plots have demonstrated that overland flow originates from them and is three times larger than in the surrounding dense shrub savannah. The next stage of land surface degradation consists of the sealed surfaces areas occurring within the open shrub savannah. As already mentioned, infiltrability within this unit is low (10 mm/hr) as assessed under simulated rainfall. In this case, rill generation is enhanced by rainsplash impact which seals the soil surface between the grass tussocks. This process is further encouraged as microtopography increases downhill. This overall system is controlled by overland flow due to impaired surface physical conditions.

As demonstrated by CHAUVEL et al. (1977) in an analogous pedologic environment in southern Senegal, pedoclimatic conditions drier than those required for ferrallitic pedogenesis, may foster the disjunction of fine and coarse soil components and trigger lateral leaching of finer particles and the transformation of ferrallitic soils into ferruginous soils. Such changes usually originate from the topsoil. This gradual deterioration of soil surface along the hillslope might be partly related to these geochemical processes.

6.2 MIDSLOPE SYSTEM

Due to the slope-break, circumstances are favourable in the midslope for the deepening of incisions. In addition, as illustrated by the example, gullies may turn into deep gullies once the incision reaches the laterally leached soils. In reducing intrasoil strength, and in precipitating gully failure events, throughflow governs the development of the deep gullies which are confined to this midslope system, whereas overland flow removes the debris from bank collapse. This system is therefore mainly controlled by a combination of processes which include increased slope angle, hence velocity, surface flow and throughflow.

As already mentioned, the slope break might have been induced by the overall lowering of land surface downstream of the midslope iron pan. This may partly result from internal lateral leaching within the layers affected by the lateral drainage of the water-table. The question still arises about the differentiation of linear erosion. Observations of deep layers located 1.5 m below a gully bottom have shown eluviation of iron along a tunnel-like feature. This suggests the existence of linear throughflow deep under the gully. Such underground flow might be responsible for piping which in turn should guide the rill pattern.

6.3 FOOTHILL SYSTEM

The foothill rills which join the stream have their own operating system, mainly controlled by the seasonal variations of the water-table level, as revealed by visual observations with piezometers. The two systems mentioned above start operating in the early rainy season and participate in loading the downslope groundwater body. At this time, except during heavy rainstorms, upstream rill flow usually infiltrates into the unsaturated sand hydromorphic soils and consequently rarely communicates with the valley bottom rills. As sand hydromorphic soils become waterlogged, due to the upward shift of the water-table, the midslope water-courses system can connect with the foot-hill system which then becomes fully operative. The hydrological connection of the three systems to the stream is therefore a time dependent-phenomenon. Long after the end of the rainy season, rill base flow may continue to drain the footslope water-table.

7 CONCLUSIONS

This study of the rill pattern within a savannah watershed leads to several conclusions:

1. The volume of linear erosion expressed per hectare is mainly governed by slope inclination. More precisely, the depth of incision is intimately related to hillslope steepness. The rill density, expressed in m/ha, is also related to slope length and topsoil clay content. However, the predicitve model of BOON & SAVAT (1981), based on simple hydraulic parameters, was validated for only 10 map units out of 18. Although slope steepness and discharge (through low permeability and slope length) actually combine to generate shear stress, they are not the only active factors. The occurrence of throughflow and the foothill water-table make the development of rill pattern more complex. When studying rill erosion in the wet tropics, one should keep in mind that surface and subsurface erosion are often combined.

2. Three rill systems can be identified: upslope incisions which include pre-rills, rills and gullies are governed by overland flow whereas the midslope system is also influenced by throughflow and piping. The initiation of deep gullies appears as a more complex process than mere deepening. The foothill rill system is

mainly controlled by the water-table level.

3. Several recurrence intervals should be taken into account in interpreting the various erosion factors. Rills may develop very quickly in a cultivated field, while the development of deep gullies may be influenced by long term pedological processes such as lateral leaching or the gradual upward encroachment of a water-table along the hillslope.

ACKNOWLEDGEMENT

The authors wish to express their sincere thanks to Prof. R. Bryan for the cheerful and constructive comments he made on the text.

8 REFERENCES

[BOON & SAVAT 1981] BOON, W. & SAVAT, J.: A nomograph for the prediction of rill erosion. In: Soil Conservation, Problems and Prospects. Ed. by R.P.C. Morgan, J. Wiley, Chichester, 303–319.

[BOULET et al. 1977] BOULET, R., BOCQUIER, G. & MILLOT, G.: Géochimie de la surface et formes du relief. I. Déséquilibre pédobioclimatique dans les couvertures pédologiques de l'Afrique tropicale de l'ouest et son rôle dans l'aplanissement des reliefs. Sciences Géologiques, **30**, 235–243.

[BROU 1985] BROU, K.: Analyse et traitement des observations des postes pluviométriques de Touba et de Bouna. Techn. Report, ORSTOM, Adiopodoume, 27 p.

[CHAUVEL et al. 1977] CHAUVEL, A., BOCQUIER, G. & PEDRO, G.: Géochimie de la surface et formes du relief: III. Les mécanismes de la disjonction des constituants des couvertures ferrallitiques et l'origine de la zonalité des couvertures sableuses dans les régions intertropicales de l'Afrique de l'ouest. Sciences Géologiques, **30**, 255–263.

[CHEVALLIER & SAKLY 1985] CHEVALLIER, P. & SAKLY, F.: Etude du rapport pluie-débit sous pluie simulée sur un petit bassin versant de savane humide (Booro-Borotou, Côte d'Ivoire). Techn. Report, Orstom, Adiopodoumé, 29 p.

[DUBREUIL 1985] DUBREUIL, P.L.: Review of field observations of runoff generation in the tropics. Journal of Hydrology, **80**, 237–264.

[GOVERS 1985] GOVERS, G.: Selectivity and transport capacity of thin flows in relation to rill erosion. CATENA, **12**, 35–49.

[KALMAN 1976] KALMAN, R.: Etude expérimentale de l'érosion par griffes. Revue de Géographie Physique et de Géologie Dynamique, **18**, 395–406.

[MOSLEY 1974] MOSLEY, M.P.: Experimental study of rill erosion. Transactions of the ASAE, **17**, 909–913.

[PLANCHON 1985] PLANCHON, O.: Utilisation d'un simulateur de ruissellement pour l'étude expérimentale de l'érosion. Tech. Report, Orstom, Adiopodoumé, 16 p.

[ROLOFF et al. 1981] ROLOFF, G., BRADFORD, J.M. & SCRIVENER, C.L.: Gully development in the deep loess hills region of central Missouri. Soil Science Society of America Journal, **45**, 119–123.

[SAVAT & DE PLOEY 1982] SAVAT, J. & DE PLOEY, J.: Sheetwash and rill development by surface flow. In. Badland Geomorphology and Piping. Ed. by R. Bryan and A. Yair. Geo. Abstracts, Norwich, 113–126.

[SAVIGEAR 1960] SAVIGEAR, R.A.G.: Slopes and hills in West Africa. Zeitschrift für Geomorphologie, Supplementband **1**, Contributions internationales à la Morphologie des Versants. Ed. by P. Birot and P. Macar, 156–171.

Address of authors:
O. Planchon, E. Fritsch and C. Valentin
ORSTOM, B.P. V-51
Abidjan, Côte d'Ivoire

RESISTANCE TO RILL EROSION: OBSERVATIONS ON THE EFFICIENCY OF RILL EROSION ON A TILLED CLAY SOIL UNDER SIMULATED RAIN AND RUN-ON WATER

R.J. **Loch**, Toowoomba and E.C. **Thomas**, Cowra

SUMMARY

Rill development under simulated rain and run-on water was studied on a grey, cracking clay on 2.87° slope regarded locally as being relatively resistant to rilling.

Three groups of plots were established under a rainulator, with all plots being 4 m wide and tilled across-slope. The first group were 3 m long, receiving rainfall only, giving a measure of rain-flow (interrill) erosion. The second group were 22 m long, receiving rainfall only, and the third group were 16.5–22 m long receiving rainfall plus run-on water. Rain was applied at 95 mm h^{-1}.

Although some scour by overland flow developed on the 22 m long plots, sediment transport by rills only developed fully on the plots receiving rainfall plus run-on, at discharges >3.0 l s^{-1}. This discharge is much larger than that found previously to cause rapid and complete development of rill erosion on two clay soils on similar slopes, under identical conditions of rainfall and runoff. There was evidence that the two soils studied previously behaved as cohesionless beds of particles, but it appears that for the soil studied in this experiment there was some cohesion within the tilled layer resisting the incision of rills until a relatively large discharge (and therefore, streampower) was reached. The critical discharge for rilling was consistent with estimated peak discharges in rill catchments for runoff events causing large field soil losses at this site.

Concentration of bed-load sediment is considered to be a better indicator of rill development than concentration of total sediment. There is potential for confusion in the identification of rill development if total sediment is used.

Methods for studying resistance to rilling are suggested.

1 INTRODUCTION

1.1 RESISTANCE TO RILLING

Several recent papers have noted that some soils are relatively resistant to rilling e.g., YOUNG & ONSTAD (1978), FOSTER et al. (1982), and YOUNG (1984). However, no detailed description of soil resistance to rilling has been pub-

ISSN 0722-0723
ISBN 3-923381-07-7
©1987 by CATENA VERLAG,
D–3302 Cremlingen-Destedt, W. Germany
3-923381-07-7/87/5011851/US$ 2.00 + 0.25

lished.

This lack of description of "resistance to rilling" is of concern for studies of erosion and soil erodibility, as the presence or absence of rills can greatly influence the results obtained. For example, MEYER et al. (1975) reported a threefold increase in soil loss when rills developed on rainulator plots. As well, presence or absence of rills can greatly alter estimates of soil erodibility derived from plots under simulated rain (LOCH 1984).

Therefore, this paper reports a study of rill development under simulated rain and run-on water on a bare, tilled soil on the eastern Darling Downs, Queensland, Australia. The soil is recognized by farmers and soil conservationists as being relatively resistant to rilling. Such resistance was confirmed by detailed measurements. The development of rills on this soil is compared with that on two soils shown by a previous study to have virtually no resistance to rilling. From this, approaches to studying both erosion on rill-resistant soils and causes of such resistance are suggested.

1.2 SEDIMENT TRANSPORT IN RILLS (OR WHEN IS A RILL NOT A RILL?)

On rough surfaces, overland flow invariably concentrates into a number of preferred flow paths (EMMETT 1978, LOCH & DONNOLLAN 1983a). It would be simple to define all lines of concentrated flow as rills. However, that approach would ignore the great changes in sediment transport capacity that can occur in such flow lines, and lead to considerable confusion and misinterpretation of the results of soil erosion studies.

FENNEMAN (1908) suggested that there were two physically distinct types of overland flow channel. Where there was net erosion, channels were incised, but where there was net deposition, the channels were shallow and lined with their own deposits of sediment. MOSS & WALKER (1978) observed the same two channel types, which they attributed to differences in bed-load transport capacity. Similarly, LOCH & DONNOLLAN (1983a) noted that flow channels were either broad and flat with considerable deposition, or, above a characteristic plot discharge, were more deeply incised, with banks undercut, and knickpoints sometimes visible. However, they noted that **those differences were not clearly visible on all plots** and therefore, some other way of assessing the erosion processes operating in the flow lines was essential.

Sediment concentrations measured by LOCH & DONNOLLAN (1983a) confirmed that sediment transport capacity differed greatly between the channel types, as suggested by MOSS & WALKER (1978). In the incised channels, concentrations of bed-load sediment were 4.80 and 4.74 times higher than in the broad flat flow lines for the two soils studied.

Such a large (and consistent) increase in capacity to transport bedload sediment suggests a marked and well-defined change in the nature of the flow in overland flow channels. Evidence for this change was reported by MOSS et al. (1982), who found that above a characteristic discharge (which was a function of bed slope), secondary flow circulations developed in sheet flow and led to the formation of incised channels. Both MOSS et al. (1979) and MOSS et al. (1982) show a large increase in bed-load transport when channelled flow developed.

The discharges reported by MOSS et al. (1982) for development of secondary flow circulations were in close agreement with the discharges noted by LOCH & DONNOLLAN (1983a) to be a threshold for high rates of bed-load transport in channel flow. Therefore, LOCH & DONNOLLAN (1983a) concluded that the two clay soils studied by them had behaved similarly to the cohesionless sand studied by MOSS et al. (1982), and that the development of secondary flow was sufficient to form incised channels with high bed-load transport capacity on those soils. (**In this paper, and in LOCH & DONNOLLAN (1983a) these incised channels are referred to as rills.**)

The two soils studied by LOCH & DONNOLLAN (1983a) differed greatly in their erodibility, whether by rain-flow (the combined action of drop impact and shallow overland flow (MOSS et al. 1979)) or rill transport. Subsequent (unpublished) studies have shown similar differences between the soils in rates of splash loss from laboratory containers. Those differences in rates of splash and rain-flow erosion did not cause any differences between the soils in rill development, although DUNNE (1980) suggested that such differences may explain presence or absence of rills in some areas. For both soils studied by LOCH & DONNOLLAN (1983a), detachment of sediment was clearly in excess of rain-flow transport capacity, with deposited sediment visible over most of the plots.

Importantly, LOCH & DONNOLLAN (1983a) also noted that rilling, on the two soils studied, developed rapidly to produce a characteristic rill sediment concentration across a wide range of discharge. In effect, the efficiency of sediment transport by channels increased sharply and reached a maximum once rilling developed. Similarly, ROSE (1985) shows evidence of a threshold streampower to be exceeded before entrainment by overland flow occurs, and of a rapid increase in efficiency of entrainment to a maximum with increasing streampower.

Erosion research frequently uses plots to obtain data on the likely behaviour of larger, field areas. If the erosion processes of importance at a field scale either do not develop, or are relatively inefficient on small plots, then results from the plots have little value.

This has particular relevance to rilling, and it is essential if studying erosion from plots to be able to determine whether sediment transport in rills has reached maximum efficiency.

2 MATERIALS AND METHODS

2.1 SOIL

The soil, a Udic Chromustert, is classified locally as a Moola clay (MULLINS, pers. comm.). It is a grey, cracking clay, formed on fine-grained sandstone, originally supporting brigalow (*Acacia harpophylla*) woodland. The 0–10 cm layer contains 10% coarse sand, 24% fine sand, 13% silt, and 53% clay. Cation exchange capacity is 37 m.e./100 g, and pH is 8.5. Organic carbon content is 1.8%, but other studies have shown no effect of organic carbon on the stability of aggregates from cracking clay soils (COUGHLAN & LOCH 1984).

2.2 SITE

The site studied is on 2.87° slope, with terraces on 0.17° slope (known locally as contour banks) installed at approximately 24 m spacing. A range of summer or winter grain crops is grown in the inter-bank area.

A detailed description of the site, its location, and of the runoff and erosion occurring under normal cropping practices is given by FREEBAIRN & WOCKNER (1986a and b).

2.3 RAINFALL SIMULATION AND PLOT PREPARATION

A rainulator, described by LOCH & DONNOLLAN (1983a), was used to apply simulated rain at 95 mm h^{-1} to plots 4 m wide, and of different lengths. (Field studies in the area have recorded 15-minute bursts of rain with intensities >80 mm h^{-1} more frequently than every second year on average (FREEBAIRN pers. comm.).) The rainulator produces drops with a median diameter of 2.1 mm, 80% of the drops being between 1.0 and 3.2 mm diameter, and a kinetic energy of rainfall of 295 kJ ha^{-1} mm^{-1} (MEYER 1958). All plots were tilled across slope with the same tined implement used by LOCH & DONNOLLAN (1983a). Rainfall intensity, preparation of the plots, measurement of runoff, and sampling and analysis of sediment followed the procedures described by LOCH & DONNOLLAN (1983a and b).

Three groups of plots were prepared, to allow study of:

(i) rain-flow erosion (MOSS et al. 1979), also termed interrill erosion;

(ii) rill erosion at "moderate" plot discharges; and

(iii) rill erosion at "high" plot discharges.

Plot length, and presence or absence of run-on water were used to control the erosion processes operating. Lengths, numbers of plots, rainfall/run-on treatments applied to obtain these three groups and plot discharges are shown in tab.1. It was not possible to standardize on a 22 m long plot because the contour bank channel would have encroached markedly on the downslope end of some plots. When applied, run-on water was added at the upslope ends of the plots, to give a 30–60% increase in plot discharge.

3 RESULTS

3.1 OBSERVATIONS

On all plots, excess rain ponded in the tillage furrows for some time before overtopping them. Runoff concentrated into flow lines, with the larger plots having only one or two flow lines per plot.

On all the longer plots, the flow lines often provided visual evidence that secondary flows had developed. The convergence of secondary flow cells resulted in a thin line of froth down the centre of the flow line. However, at discharges <3.0 l s^{-1} approximately, these flow lines were subsequently found to be relatively broad and not incised (phot.1). However, while rain and runoff were in progress, it was impossible to determine visually the extent to which sediment transport in rills approached maximum efficiency.

Treatment	Plot length (m)	No. of plots	Maximum plot discharge ($l\ s^{-1}$)
Rainfall only	03.0	4	<0.3
Rainfall only	22.0	2	2.5
Rainfall plus run-on	16.5	1	4.2
	21.0	1	4.0
	22.0	1	3.3

Table 1: *Plot lengths, numbers of plots, and treatments applied.*

Photo 1: *Flow line on a 22 m long plot that did not receive run-on water, showing the lack of incision at discharges $<3.0\ l\ s^{-1}$. Flow direction is from right to left.*

3.2 DISTINCTION BETWEEN BED AND SUSPENDED LOADS

For the flow conditions applying under the rainulator, 20 μm is taken as an approximate boundary between bed and suspended loads. This division is based on apparent velocities of sediment movement. A comprehensive explanation of this approach is given by MOSS et al. (1979). For a plot 3 m long, LOCH & DONNOLLAN (1983b) noted that sediment in the size range 20–125 μm moved relatively slowly, which is consistent with movement by saltation (MOSS et al. 1979). (Both suspended load and contact bed load moved much more quickly.) Therefore, there is justification for suggesting that sediment >20 μm moved as bed-load on similar 3 m long plots in this experiment.

However, at the higher discharges developed on longer plots, it could be questioned whether a separation of bed and suspended loads at 20 μm remains valid. Nor is assessment of relative velocities

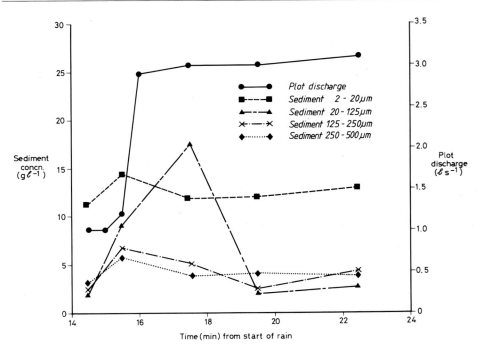

Figure 1: *Changes through time in plot discharge and concentrations in runoff of several sediment size fractions for a 21 m long plot on the Moola clay.*

Mean concentrations in g l^{-1} of:	Plot grouping		
	3 m long	22 m long, without run-on, for discharges > 0.6 l s^{-1}	16.5–22 m long, plus run-on, for discharges > 3.0 l s^{-1}
Total sediment	16.8	29.4	50.4
Bed-load sediment	9.4	23.3	42.0

Table 2: *Effects of plot size and discharge on concentrations of total and bed-load sediment (>20μm).*

of sediment movement generally possible on the longer plots, as overtopping of surface storage and development of surface runoff tends to be more gradual than on the 3 m long plots. It is therefore difficult to find a point in the hydrograph where large quantities of sediment would have begun moving at the one time, and at relatively high discharges.

Fortunately, one 21 m long plot did show a sharp increase in runoff and a corresponding increase in sediment concentration. Changes in sediment size fractions through time (fig.1) showed the 20–125 μm size fraction of sediment to be slower to peak than the two larger size fractions (0.125–0.250, and 0.25–0.50 mm). From this it can be inferred that sediment 20–125 μm moved relatively slowly, and therefore, largely by saltation. Consequently, it seems reasonable to regard sediment >20 μm as

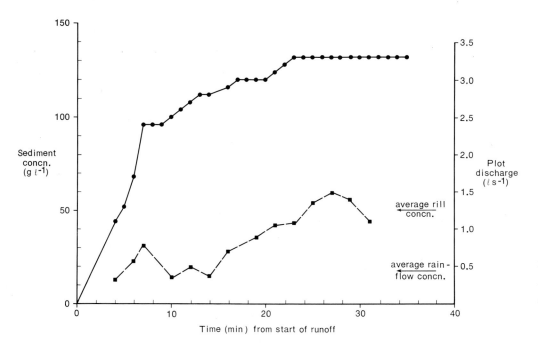

Figure 2: *Sediment concentrations and plot discharges from a 22 m long plot on the Moola clay, under simulated rain and run-on water. Average sediment concentrations (over all plots) for rill flow (for discharges >3.0 l s^{-1}) and for rain-flow transport (from the 3 m long plots) are shown.*

bed-load for all plots. However, further study of sediment transport mechanisms is needed to confirm this conclusion.

3.3 SEDIMENT CONCENTRATIONS

Average concentrations of total sediment and sediment >20 μm are shown in tab.2.

For the group of longer plots receiving rainfall only, sediment concentrations were averaged over the time period when discharges were greater than 0.6 l s^{-1}, but for the group of longer plots receiving rainfall plus run-on, sediment concentrations were averaged for discharges greater than 3.0 l s^{-1}.

The discharge of 0.6 l s^{-1} was selected because rills formed at this discharge on rill-susceptible soils on 2.29° slope (LOCH & DONNOLLAN 1983a). If the Moola clay (which is on 2.87° slope) was similarly susceptible to rilling, then it could also be expected to rill at this discharge. However, sediment concentrations for discharges >0.6 l s^{-1} were not consistent with rill transport being fully developed (tab.2). The discharge of 3.0 l s^{-1} was selected as being that at which sediment concentrations showed evidence of rilling (fig.2).

4 DISCUSSION

4.1 DEGREE OF DEVELOPMENT OF RILLS

Previous rainulator studies with soils susceptible to rilling have shown sediment transport in rills to reach maximum efficiency at relatively low discharges (LOCH & DONNOLLAN 1983a). Results from the Moola clay did not follow this pattern. Instead, increasing plot discharge caused a much more gradual increase in sediment concentration (fig.2), indicating that the discharges needed for rill transport to reach maximum efficiency were much larger than for rill-susceptible soils.

One way to assess whether sediment transport in rills has reached maximum efficiency is to compare concentrations of bed-load sediment. This approach was used by LOCH & DONNOLLAN (1983a). For two clay soils on 2.29° slope they found that rills carried 4.74 and 4.80 times more bed-load sediment than rain-flow. As noted previously, they suggested that this constant order of difference was a measure of the different bed-load transport capacity of rills and rain-flow, as a result of the development of secondary flow circulations in rills. Assuming that the slight differences in slope (2.29° slope in the work of LOCH & DONNOLLAN (1983a), 2.87° slope for the Moola clay) would not greatly alter this relativity, a similar order of difference in bed-load transport between rills and rain-flow could be expected for the Moola clay.

For the plots receiving run-on, average concentrations of bed-load at discharges >3.0 l s^{-1} were 4.5 times the average concentrations of bed-load in runoff from the 3 m long (rain-flow eroded) plots. This is encouraging agreement with the data of LOCH & DONNOLLAN (1983a), suggesting that rill transport on the Moola clay had nearly reached its maximum efficiency.

Tab.2 shows further justification for assessing rill development on the basis of bed-load sediment. For all three groups of plots, concentrations of suspended sediment are similar (6.1–8.4 gl^{-1}), with the only difference between groups of plots being in bed-load sediment.

For well-aggregated clay soils, aggregate breakdown to $<20\mu$m is due largely to raindrop impact (LOCH & DONNOLLAN 1983b), so some dilution of suspended load could have been expected on plots where run-on water was added. However, it appears that such dilution was balanced by further breakdown of aggregates to $<20\mu$m in rill flow. Similarly, LOCH & DONNOLLAN (1983b) found evidence that breakdown of aggregates to $<20\mu$m occurred in rill flow on a similar soil, though raindrop impact was the main cause of such breakdown. The amounts of sediment $<20\mu$m produced in rills in this experiment must have been sufficient to cancel out the 30–60% dilution caused by run-on. Interestingly, the 22 m long plots not receiving run-on showed no increase in concentrations of sediment $<20\mu$m relative to the 3 m long plots, even though some erosion by overland flow had developed. It appears that for well-aggregated soils the onset of entrainment by overland flow is characterised by increased transport of bed-load sediment, with breakdown of sediment to $<20\mu$m only occurring when the stronger turbulence of rilling is fully established.

4.2 COMPARISON WITH FIELD MEASUREMENTS

Runoff from the contour bay and soil movement into the contour bank channel were measured at this site over a 6-year period (FREEBAIRN & WOCKNER 1986a). Examination of their data shows that for individual runoff events, appreciable soil losses (>5.5 t ha^{-1}) from the inter-bank area and a relatively high ratio of soil loss to runoff occurred only when peak runoff rate at the catchment outlet was greater than 12 mm h^{-1}.

Rills in the contour bay catchment have an average spacing of approximatively 30 m, so that the contour bank effectively drains a number of rill catchments. A study of slightly larger contour bays showed that peak runoff rate for a rill catchment was 2.6 times that measured at the contour bay outlet for one runoff event (CIESIOLKA & FREEBAIRN 1982). From that study we assume that at the Moola clay site the rill catchments (20 m long and 30 m wide) would have had peak runoff rates approximately twice those of the contour bay catchment. Therefore, a runoff rate greater than 24 mm h^{-1}, giving a peak discharge of at least 4.0 l s^{-1} would have been required in the rill catchments before runoff events caused appreciable erosion. This is in good agreement with the rainulator data showing a discharge >3.0 l s^{-1} needed for rilling on this soil to reach full efficiency.

4.3 RESISTANCE TO RILLING — PARTICLE SIZE OR COHESION?

For rill-susceptible soils, LOCH & DONNOLLAN (1983a) suggested that the development of secondary flow circulations was sufficient to cause rilling. That is clearly not the case for the rill-resistant Moola clay.

It should be noted that the 22 m long plots receiving rainfall only gave much higher sediment concentrations than the 3 m long, rain-flow eroded plots (tab.2). This is evidence that some entrainment by overland flow must have developed, and after rain and runoff ceased, scouring of the soil surface was clearly visible in the flow lines (phot.1). It seems that resistance to rilling on this soil is not a general resistance to entrainment by overland flow, but rather, a resistance to the incision of channels to form rills.

DUNNE (1980) suggested that incision of rills would be prevented if the rate of sediment transport to rills (by rainsplash and sheetwash) exceeded the transport capacity of the rills. If that explanation applied to the Moola clay, sheetwash and rainsplash from that soil would be greater than from either Irving clay or Middle ridge clay loam. Instead concentrations of sediment carried by rain-flow transport under identical rainfall and runoff were higher for the Irving clay (22 g l^{-1} compared with 16.9 g l^{-1} for the Moola clay). LOCH & SMITH (1986) report that amounts of soil splashed from small containers under a laboratory rainfall simulator were almost identical for Moola (grey) and Irving clays. Therefore, other explanations of the relative resistance of the Moola clay to rilling must be considered.

Resistance to scour has been commonly related to particle size (GRAF 1971). However, this possibility is not consistent with measured sediment sizes. For the Moola clay, the median size of sediment carried by rills at discharges >3.0 $l\ s^{-1}$ was in the range of 0.25–0.125 mm. In comparison, median sizes of sediment in rill flow from the two soils susceptible to rilling studied by LOCH & DONNOLLAN (1983a) were approximately 0.5 mm. Therefore, it is more likely that aggregates in the tilled layer of the Moola clay were finer than in the tilled layer of the two soils susceptible to rilling.

From this it could be suggested that cohesion in the soil mass would have been greater in the Moola clay, and is a more likely explanation of its resistance to rilling. Certainly, the delayed onset of rilling (but complete development of rills provided a large enough discharge is reached) is what would be expected if cohesion of the tilled layer had to be overcome before rills could form. In terms of defining cohesion of the soil mass, it would seem most useful to focus on ways of measuring the physical state of the tilled layer under rain, **at the water contents at which erosion occurs**. Various types of shear strength measurements could be considered, but generally shear vanes or similar instruments do not seem to be well-suited to wet, cloddy tilled soils.

A further complication is that the tilled layer is not homogeneous, having a relatively fine, compacted layer at the surface (due to breakdown and compaction by raindrops), and coarser material below, so there is the problem of identifying the layer actually preventing incision of rill channels.

It could also be expected that cohesive soils would show greater variability in erosion from plot to plot for a given treatment. There is also the possibility that tillage (or lack of it) may govern whether, and to what extent, a soil is relatively cohesive (FOSTER et al. 1982). In terms of management to reduce erosion of agricultural lands, these considerations may be extremely important.

4.4 APPARENT RESISTANCE TO RILLING AND SEDIMENT SIZE

The development of secondary flow circulations and incision of rills results in a large increase in the capacity of overland flow to transport sediment. However, that increased transport capacity will only increase sediment transport if transport capacity is limiting. Sediment supply may be limiting because of a lack of detached sediment, as shown by ALBERTS et al. (1980). Alternatively, much of the sediment available may be so fine that it is carried in suspension and the capacity of overland flow to carry suspended sediment would be unlikely to be exceeded.

For aggregated clay soils, raindrop impact appears to be a major factor in the breakdown of aggregates to suspendable sizes (LOCH & DONNOLLAN 1983b). The supply of suspended sediment would therefore be largely controlled by raindrop impact. If a soil is particularly unstable under rain, the rate of supply of suspended load may be quite high, and bed-load would become a minor component of the total sediment load. Under such conditions, even a four-fold increase in the rate of bed-load transport may not greatly increase total sediment transport, and thus, rilling would **appear** to be ineffective. Consistent with this, YOUNG (1984) found that if aggregate breakdown under rain was particularly high, the soils were "resistant to rilling". However, it is probably misleading to refer to much soils as rill-resistant. It may be more useful to identify such soils as being ones which erode dominantly as suspended material and on which the movement of soil by rilling is relatively less important.

4.5 METHODS FOR STUDYING EROSION ON SOILS RESISTANT TO RILLING

Even on soils susceptible to rilling, studies of rilling can encounter problems. LOCH (1984) discussed interactions of plot size and slope steepness, and noted that if rilling is to be studied, then plots must be large enough for plot discharge to exceed that needed to form rills at the slope steepness studied.

Assuming that critical plot discharge/slope steepness interactions are taken into account, there remain the problems of identifying and characterizing rill-resistant soils.

To identify rilling (or its absence), some basis for comparison is essential. Logically, this would be a measure of rain-flow erosion. The absence of rilling can be ensured by using a relatively short plot (3 m long in this study). However, it should be noted that there can be differences between rainfall simulators (GLANVILLE 1984), so it is important that a standardized, and well-understood measure of rain-flow be adopted.

As a basis for comparing rain-flow transport with that from a supposedly rilled plot, total sediment concentration is inappropriate, but concentration of bed-load sediment is quite effective. However, the ratio between rill and rain-flow concentrations of bed-load sediment noted by LOCH & DONNOLLAN (1983a) (and used in this paper), would almost certainly alter for different slopes or rainfall simulators, and further study is needed to allow more general application of this observation.

Nonetheless, the approach outlined gives some potential to assess:

(a) whether the plots established **could** have rilled; and

(b) whether rilling **did** fully develop.

One further aspect is essential—provision to run plots at larger discharges if initial examination of results shows that sediment transport by rills did not develop to maximum efficiency.

As this paper shows, it would be wrong to assume that because rills did not develop or were relatively inefficient on a plot at a given (low) discharge, rills would be similarly absent or inefficient at larger discharges. Studies of rill erosion or of resistance to rilling using small plots and low discharges are likely to be of dubious value at best.

5 CONCLUSIONS

Further study of soil resistance to rilling is needed. Not only to provide data on the discharges (or preferably, streampowers or tractive forces) at which rills form on cohesive soils, but also to seek ways of measuring cohesion of the tilled layer so that critical discharges for rilling can be predicted. The potential variability of erosion on rill-resistant soils may add to the difficulty of such studies, but may also encourage study of those aspects of tillage influencing resistance to rilling.

ACKNOWLEDGEMENT

We thank Mr and Mrs W. Bosse for allowing the experiment to be carried out on their property, and Mr D. Ashcroft for his assistance with the operation of the rainulator.

6 REFERENCES

[ALBERTS et al. 1980] ALBERTS, E.E., MOLDENHAUER, W.C. & FOSTER, G.R.: Soil aggregates and primary particles transported in rill and interrill flow. Soil Science Society of America Journal **44**, 590–595.

[CIESIOLKA & FREEBAIRN 1982] CIESIOLKA, C.A.A. & FREEBAIRN, D.M.: The influence of scale on runoff and erosion. In: Resources — Efficient Use and Conservation. Agricultural Engineering Conference, Armidale, NSW, Australia, The Institution of Engineers, Australia, National Conference Publication No. **82/8**, 203–206.

[COUGHLAN & LOCH 1984] COUGHLAN, K.J. & LOCH, R.J.: The relationship between aggregation and other soil properties in cracking clay soils. Australian Journal of Soil Research **22**, 59–69.

[DUNNE 1980] DUNNE, T.: Formation and controls of channel networks. Progress in Physical Geography **4**, 211–239.

[EMMETT 1978] EMMETT, W.W.: Overland flow. In: Hillslope Hydrology. (Ed. M.J. Kirkby), John Wiley and Sons, New York, 145–176.

[FENNEMAN 1908] FENNEMAN, N.M.: Some features of erosion by unconcentrated wash. Journal of Geology **16**, 746–754.

[FOSTER et al. 1982] FOSTER, G.R., JOHNSON, C.B. & MOLDENHAUER, W.C.: Critical slope lengths for unanchored cornstalk and wheat straw residue. Transactions of the ASAE **25**, 935–939 and 947.

[FREEBAIRN & WOCKNER 1986a] FREEBAIRN, D.M. & WOCKNER, G.H.: A study of soil erosion on vertisols of the eastern Darling Downs, Queensland. I. the effect of surface conditions on soil movement within contour bay catchments. Australian Journal of Soil Research **24**, 135–158.

[FREEBAIRN & WOCKNER 1986b] FREEBAIRN, D.M. & WOCKNER, G.H.: A study of soil erosion on vertisols of the eastern Darling Downs, Queensland. II. The effect of soil, rainfall and flow conditions on suspended sediment losses. Australian Journal of Soil Research **24**, 159–172.

[GLANVILLE 1984] GLANVILLE, S.F.: Rotating disc rainfall simulator. In: Soil Erosion Research Techniques. Conference and Workshop Series, No. QC84001. Queensland Department of Primary Industries, Brisbane, 10–18.

[GRAF 1971] GRAF, W.H.: Hydraulics of Sediment Transport. McGraw-Hill.

[LOCH 1984] LOCH, R.J.: Field rainfall simulator studies on two clay soils of the Darling Downs, Queensland. III. An evaluation of current methods for deriving soil erodibilities (K factor). Australian Journal of Soil Research **22**, 401–412.

[LOCH & DONNOLLAN 1983a] LOCH, R.J. & DONNOLLAN, T.E.: Field rainfall simulator studies on two clay soils of the Darling Downs, Queensland. I. The effects of plot size and tillage orientation on erosion processes and runoff and erosion rates. Australian Journal of Soil Research **21**, 33–46.

[LOCH & DONNOLLAN 1983b] LOCH, R.J. & DONNOLLAN, T.E.: Field rainfall simulator studies on two clay soils of the Darling Downs, Queensland. II. Aggregate breakdown, sediment properties and soil erodibility. Australian Journal of Soil Research **21**, 47–58.

[LOCH & SMITH 1986] LOCH, R.J. & SMITH, G.D.: Measuring aggregate stability under rainfall. In: Assessment of Soil Surface Sealing and Crusting. (Ed. F. Callebaut, D. Gabriels and M. de Boodt), Flanders Research Centre for Soil Erosion and Soil Conservation. 146–153.

[MEYER 1958] MEYER, L.D.: An investigation of methods for simulating rainfall on standard runoff plots and a study of the drop size, velocity and kinetic energy of selected spray nozzles. Special Report 81, Soil and Water Conservation Research Division, Agricultural Research Service, U.S. Department of Agriculture.

[MEYER et al. 1975] MEYER, L.D., FOSTER, G.R. & ROMKENS, M.J.M.: Source of soil eroded by water from upland slopes. Proc. 1972 Sediment Yield Workshop, US Dept. Agric. Sediment. Lab., Oxford, Mississippi, ARS-S-40, USDA, 177–189.

[MOSS & WALKER 1978] MOSS, A.J. & WALKER, P.H.: Particle transport by continental water flows in relation to erosion, deposition, soils and human activities. Sedimentary Geology **20**, 81–139.

[MOSS et al. 1979] MOSS, A.J., WALKER, P.H. & HUTKA, J.: Raindrop — stimulated transportation in shallow water flows: an experimental study. Sedimentary Geology **22**, 165–184.

[MOSS et al. 1982] MOSS, A.J., GREEN, P. & HUTKA, J.: Small channels: their experimental formation, nature and significance. Earth Surface Processes and Landforms **7**, 401–415.

[ROSE 1985] ROSE, C.W.: Developments in soil erosion and deposition models. In: Advances in Soil Science, Volume **2** (Ed. B.A. Stewart), Springer-Verlag, New York, 1–64.

[YOUNG & ONSTAD 1978] YOUNG, R.A. & ONSTAD, C.A.: Characterization of rill and inter-rill eroded soil. Transactions of the ASAE **25**, 1126–1130.

[YOUNG 1984] YOUNG, R.A.: A method of measuring aggregate stability under waterdrop impact. Transactions of the ASAE **31**, 1351–1354.

Addresses of authors:
R.J. Loch
Queensland Wheat Research Institute
P.O. Box 5282
Toowoomba, Australia 4350
E.C. Thomas
Soil Conservation Research Centre
P.O. Box 445
Cowra, Australia 2794
formerly: Queensland Department of Primary Industries
P.O. Box 102
Toowoomba, Australia 4350

RILL EROSION ON ARABLE LOAMY SANDS IN THE WEST MIDLANDS OF ENGLAND

M.A. **Fullen** & A.H. **Reed**, Wolverhampton

SUMMARY

Rill erosion on arable loamy sands is examined at Hilton, east Shropshire. Erosion rates were measured on an array of ten 25 m^2 runoff plots maintained in a bare condition on 7–15° slopes. Rills usually developed during brief, high intensity (>10 mm hr^{-1}) convective summer storms on lower to middle sections of plots and were largest and most extensive on slopes $>13°$. In subsequent storms rills deepened and extended upslope by headward erosion. Rills measured 1–8 cm deep and 1–10 cm wide. Plot observations suggest that very small catchment areas can maintain rill growth.

Agricultural operations significantly affect rill development, as tractor wheelings and drill lines often act as initial axes of erosion, which are incised into the Ap horizon during erosion episodes. The number of tractor passes, orientation of wheelings in relation to slope and soil moisture conditions during field operations are important factors influencing the extent and severity of rill erosion along cultivation lines.

1 INTRODUCTION

Water erosion of arable soils is increasingly recognised as a problem in Britain (REED 1979a,b, EVANS 1980, FULLEN 1985a). Prior to crop emergence fallow soils are vulnerable to erosive rains, which can initiate rill systems. Incision of rill networks can result in gully erosion, causing loss of large quantities of topsoil (DOUGLAS 1970, EVANS & NORTCLIFF 1978, REED 1979b). Current agricultural practices, such as field enlargement, hedgerow removal, cultivation of steep slopes, soil compaction and runoff concentration within cultivation lines may be accelerating erosion (REED 1979b, FULLEN 1985a).

Investigations in east Shropshire show that rill erosion is significant on arable soils, but its importance varies spatially and temporally. It is intimately related to agricultural operations. Tractor tyres and seed drills imprint surface channels, which can act as rill systems during erosive rains. This paper examines the relationships between soil management and rill erosion.

ISSN 0722-0723
ISBN 3-923381-07-7
©1987 by CATENA VERLAG,
D–3302 Cremlingen-Destedt, W. Germany
3-923381-07-7/87/5011851/US$ 2.00 + 0.25

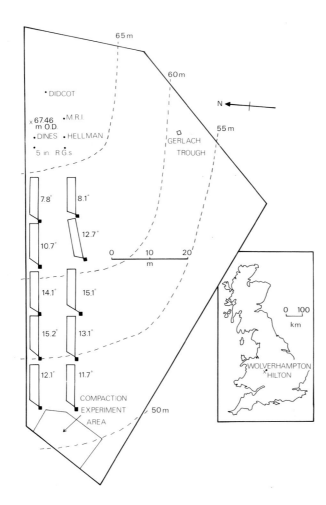

Figure 1: *Location and plan of Hilton field site.*
KEY: Sat A plots left, B right, 1 top, 5 bottom.
Didcot: Didcot 9 channel automatic weather station.
M.R.I.: Meteorological Research Incorporated automatic weather station.
Dines: Dines autographic rain gauge.
Hellman: Hellman autographic rain gauge.
5 in. R.G.s: Standard British Meteorological Office 5 inch (127 mm) diameter rain gauges.

2 FIELD STUDIES AND METHODS

Field studies were mainly within an area of approximately 100 km^2 in east Shropshire, in the West Midlands of England. The local Permo-Triassic sandstones carry sandy loam and loamy sand soils of the Newport and Bridgnorth Associations (SOIL SURVEY OF ENGLAND AND WALES 1983). Their light, porous, freely drained and easily cultivated nature make these soils suitable for arable use; the most common crops are cereals (barley, wheat) and root crops (sugar beet, potatoes). But the intensification of arable cultivation has contributed to a serious soil erosion problem in the area (REED 1979a, b, FULLEN 1984, 1985b, FULLEN & REED 1986).

Measurements were made at a field station near Hilton. Rainfall parameters (amount, duration, intensity) were measured using a suite of non-recording and autographic rain gauges and two automatic weather stations (fig.1). Ten

25 m² runoff plots were constructed near the Hilton meteorological site on slopes ranging from 7 to 15°. Plot topsoils were slightly to moderately stony loamy sands of the Bridgnorth series. The mean particle-size distribution of 20 samples from the plots was sand (2000–63 µm) 79.8% by weight ($\pm$ S.D. 7.0), silt (63–2 µm) 14.8% ($\pm$ 4.6) and clay (<2 µm) 5.4% ($\pm$ 3.1). Eight plots operated simultaneously from 1 June 1982. The lower plots (A5, B5) became operational in October 1982. Runoff and erosion rates were regularly measured until September 1984 (FULLEN & REED 1986). A 0.5 m wide Gerlach trough was installed on a bare 12° slope (fig.1).

The effect of farming operations and machinery on soil structure, hydrology and erosion were investigated using various complementary techniques. Soil structural condition was evaluated using a Vicksberg penetrometer and soil bulk density measurements. Infiltration rates were measured with double-ring flooding infiltrometers (FULLEN 1985b). Soil organic content was determined by loss-on-ignition (850°C for 30 min).

3 RESULTS

3.1 RILL EROSION ON BARE ARABLE SOILS

To simulate seed-bed conditions without the added effects of mechanical compaction, plot soils were mechanically rotovated and raked level to produce a friable porous tilth. Plot soils were then allowed to compact by raindrop impact and splash. The low organic content ($\sim$ 2%) and uncemented medium granular structure of soil aggregates allowed moderate falls of rain (10–15 mm) to cause rapid slaking of the surface and the formation of single-grained caps 1–4 mm thick.

The erodible nature of these capped soils is illustrated by cumulative annual soil losses >4 kg m^{-2} for the eight plot array (A1–A4, B1–B4) (tab.1). However, erosion rates were very variable and closely related to the contribution of rill erosion relative to other soil transporting mechanisms (i.e. splash and sheetwash erosion). Rill erosion was particularly important during summer convective storms, and greatest on steep slopes.

Most erosion was accomplished by brief, high intensity convective storms. Between 1 June 1982 and 1 June 1984 five convective storms, with intensities >10 mm hr^{-1}, accounted for 89.6% of erosion on the eight plot array. During 27 months of plot monitoring six short-duration convective storms accounted for 82.7% of total erosion. During summer (June, July, August) 1984 72% (854 kg) of the erosion was caused by a convective storm on 14 August, during which 30.7 mm of rain fell at a peak intensity of 20 mm in 30 minutes. The erosive storms were short. On 28 April 1983 11.6 mm fell in 40 minutes and caused 492 kg of plot erosion and on 18 July 1983 a 11.2 mm 15 minute fall resulted in 599 kg of erosion.

Rills were incised into the Ap horizon during intense storms and formed the main sediment routes to the plot outlet. They acted both as receptors of material splashed or washed from inter-rill areas and as relatively efficient channels of concentrated runoff. For instance, a small rill, 2–3 cm deep, 3 cm wide and extending some 5 m upslope, was incised to the Gerlach trough during the 28 April 1983 storm (fig.1). By direct incision into the bare soil and by conducting sheetwash in its catchment area, this

Variable	1 June 1982–1 June 1983.	1 June 1983–1 June 1984.	1 June–31 August 1984.
Total rainfall (mm)	749.8	509.1	148.5
Total soil loss (kg)	937.0	836.4	1180.0
Soil loss equiv. (t ha^{-1}y^{-1})	46.9	41.8	59.0
Total soil loss (kg m^{-2})	4.7	4.2	5.9
Surface lowering (mm)*	3.8	3.4	4.8

* Based on mean capped soil bulk density of 1.24 g cm^{-3} (± 0.09, N = 50) (cf. tab.4)

Table 1: *Erosion on the eight bare plots at Hilton June 1982–August 1984.*

Plot	Slope (Degrees)	1 June 1982–1 June 1983 Erosion (kg)	1 June 1982–1 June 1983 Erosion equiv. (t ha^{-1})	1 June 1983–1 June 1984 Erosion (kg)	1 June 1983–1 June 1984 Erosion equiv. (t ha^{-1})
A1	7.8	42.7	17.1	12.3	4.9
A2	10.7	28.0	11.2	34.9	14.0
A3	14.1	122.2	48.9	163.9	65.6
A4	15.2	171.2	68.5	144.6	57.9
B1	8.1	56.1	22.4	30.7	12.3
B2	12.7	114.8	45.9	121.7	48.7
B3	15.1	183.4	73.4	167.8	67.1
B4	13.1	218.9	87.6	160.3	64.1
A5	12.1	-	-	6.82	2.7
B5	11.7	-	-	18.42	7.4

Table 2: *Plot erosion in relation to slope.*

rill appeared to contribute virtually all the 70.4 kg of sediment received by the Gerlach trough collector. These results agree with field observations on Belgian loamy sands (GABRIELS et al. 1977). YOUNG & WIERSMA (1973) tried to distinguish the relative contributions of rill and inter-rill erosion, and suggested that some 80% of sediment eroded during laboratory simulations was by rills. Our observations suggested that proportion at Hilton was of the same order of magnitude. Rill runoff was capable of transporting large particles; during the 28 April 1983 storm a 35 cm^3-105.9 g pebble was removed from plot B4.

After intense rains a dendritic pattern of rills was left on plot soils (phot.1). Rills were 1–8 cm deep and 1–10 cm wide. In subsequent storms rills both deepened and extended upslope by headward retreat, in a manner comparable to the laboratory study of MOSLEY (1974). For example, rills incised during the storms of summer 1982 were extended through 27.9 mm of rain between 24–27 September 1982 by 0.70 m on plot B3 and 1.09 m on plot B4, and eventually incised to within 1.50 and 0.89 m respectively of the plot headboards. This suggests that small catchment areas can maintain rill growth, which accords with conclusions from laboratory simulations by DE PLOEY (1981). In a rilled state plots were highly erodible; for instance, the 14 August 1984 storm left plot soils in a highly rilled condition and a subsequent rainfall of only 4.9 mm on 23 August resulted in another 22.9 kg of plot erosion. However, rills tended to be ephemeral features, because low-intensity rain caused splash erosion to obliterate them by sediment choking.

Rill development was highly responsive to slope angle. Tab.2 shows the marked acceleration of erosion with slope, increasing four-to-five fold with a doubling of slope angle. Following high intensity storms rills were especially large and extensive on slopes >13° (A3, A4, B3, B4).

Plot	28 April 1983	18 July 1983
A1	13.3	7.4
A2	13.3	28.7
A3	76.2	124.3
A4	96.7	109.3
B1	20.0	23.8
B2	64.1	98.5
B3	88.4	102.4
B4	120.0	104.3
A5	0.26	0.72
B5	6.8	6.0

Table 3: *Plot erosion (kg) resulting from two intense storms.*

Soil organic content appeared to affect degree of rill development. Plots A5 and B5 became operational in October 1982 immediately after rotovation from pasture and so contained more organic matter ($\overline{X}$ 3.95% by weight, $\pm$ 0.40, N=20) than other plots, which had been bare for five years (e.g. plots A1, B1 mean organic content 2.55%, $\pm$ 0.34, N=20). Initially these plots were considered as an extention of the pre-existing eight plot array but their erosional behaviour was so different that results were not considered comparable. Soils in plots A5 and B5 had better aggregate structure, were less prone to capping and, despite being on strongly sloping to moderately steeply sloping sites, experienced low erosion rates (tab.2). Indeed, rilling occurred only infrequently and networks tended to be shallow and weakly developed. Tab.3 shows the relatively low erosion rates experienced by plots A5 and B5 during two erosive storms.

3.2 AGRICULTURAL OPERATIONS AND RILL DEVELOPMENT

Movement of agricultural machinery over cultivated soil causes soil compaction, resulting in increased bulk density (TROUSE 1966, SOANE et al. 1980), loss of transmission and storage pores (GREENLAND 1977), and a reduction in infiltration rate (DE HAAN & VAN DER VALK 1970, GAHEEN & NJØS 1977, LINDSTROM & VOORHEES 1980). Rainfall then has a greater propensity to run off slopes rather than infiltrate into the solum, thus increasing the likelihood of water erosion. The compaction effect is exacerbated if cultivation practices provide efficient conduits for concentrated runoff in the form of tractor wheelings and drill lines. These channels can act as initial axes of erosion, which become incised into the Ap horizon during repeated erosion episodes.

In east Shropshire compaction by farm machinery increased soil bulk density (tab.4), and decreased infiltration rates. Infiltration rates on capped soils were fairly high (tab.5), but infiltration into wheelings formed by a 4.5 t tractor were more than two orders of magnitude less (tab.5). These measurements were generally applicable because ponded water frequently remained in wheelings for several days, indicating rates were <1 mm hr^{-1}

Diminished infiltration is already known to increase the tendency for slope runoff, even at low rainfall intensities. REED (1983) and FULLEN (1985b) considered intensities as low as 1 mm hr^{-1} to be erosive on moist, compacted bare soils in Shropshire. Soil compaction and up and down slope cultiva-

Soil condition	Max.	Min.	$\overline{X}$	S.D.	N	t value*
Capped	1.45	1.01	1.24	0.09	50	-8.43
Artificially compacted	1.71	1.12	1.41	0.11	50	

* Pooled t value $P < 0.001$, d.f. = 98.

Table 4: *Bulk density values of east Shropshire soils ($g\ cm^{-3}$).*

Condition	Max.	Min.	$\overline{X}$	S.D.	N
Crusted	78.60	5.15	30.16	17.85	88
Tractor wheeling	0.37	0.016	0.13	0.07	126

Table 5: *Infiltration rate measurements on east Shropshire soils ($mm\ hr^{-1}$).*

Photo 1: *Rill network on plot B4 after 28 April 1983 storm.*

Photo 2: *Confined rill erosion, Wrottesley Lodge Farm, May 1983.*

Photo 3: *Unconfined rill erosion, Trescott, March 1984.*

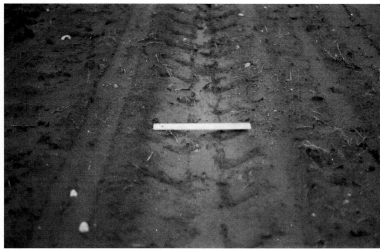

Photo 4: *Cutting of a rill through a tractor wheeling, Boningale, April 1983.*

tion lines were considered as major contributory factors at over 1000 water erosion sites identified in the West Midlands between 1965 and 1983 (REED 1983). The concentration of rill erosion on cultivation lines has been termed "confined rill erosion" by REED (1979a, b), (e.g. phot.2). Where rills break out of cultivation lines the erosion can be described as "unconfined rill erosion" (phot.3). In extreme cases rill development can contribute to gully formation (confined or unconfined) (REED 1979a,b).

The number of tractor passes has an important bearing on rill development. REED (1983) noted that deeper rill features frequently formed in wheelings subjected to two or more passes, while adjacent wheelings produced by one tractor pass and drill lines received splashed ma-

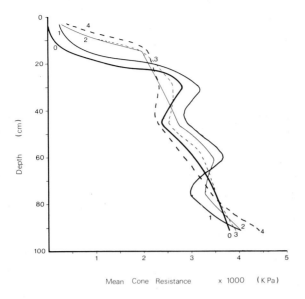

Figure 2: *Penetrographs through soils on the Hilton site subject to different numbers of tractor passes (mean of 10 penetrations).*

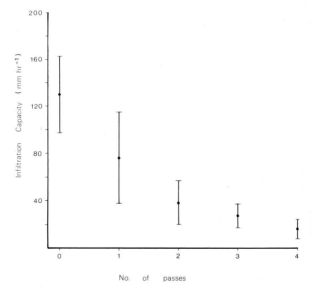

Figure 3: *Mean infiltration capacity (mm hr^{-1}) ($\pm$ 1 S.D., N=50) versus number of tractor passes.*

terial or developed smaller rills. Multiple tractor passes compress the soil further and decrease infiltration rates further, making rill erosion on these wheelings more likely. The compressive effect of tractor wheels is greater on moist soils, especially at field capacity (AKRAM & KEMPER 1979). Hence, multiple tractor passes on moist sandy soils are likely to encourage even greater rill erosion.

To examine the hydrological effects of varying tractor passes a 50 m^2 area of the experimental site was cleared, rotovated and raked level and the 4.5 t tractor produced wheelings by one, two, three or four passes (fig.1). Two infil-

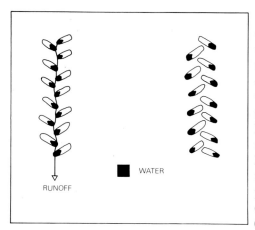

Figure 4: *Sketch of effect of tractor lug imprints on surface water movement. Ponding of water in centre of lugs encourages integration and rill initiation. Ponding in extremities of lugs impedes integration.*

trometers were inserted into each set of tractor wheelings and into uncompacted soil. Infiltration capacity was measured on each of the ten infiltrometers after two hours saturation. Ten penetrographs were also taken on each of the five treatments. The experiment was carried out in May 1984 when the soils were dry ($\overline{X}$ 7.18% moisture by weight 0–5 cm deep, ± 0.98, N = 10).

Fig.2 shows that in dry topsoil compaction increased slightly with the number of tractor passes; it was effective only to ~15–20 cm depth, with no systematic trends apparent at greater depth. Despite only moderate shallow compaction infiltration capacity was considerably decreased on the wheelings, especially after two or more passes (fig.3). Changes in structure and infiltration properties on soils compacted in varying moisture conditions are receiving further investigation.

The direction of tractor movement up or down slope is often important in the development of confined rill systems. REED (1979a) noted that deeper confined rills developed in wheelings where the apex of the V-shaped lug pattern of the tyre was oriented downslope (i.e. where tractor movement was upslope). Water accumulates in the lug depressions, which overtop during rainfall events and combine with other ponded depressions. Eventually a rill incises through the centre of the wheeling (phot.4). In contrast, where the V-shaped pattern is orientated upslope water tends to be more diffuse and thus rill erosion is less likely (fig.4). The importance of tractor wheeling as sediment sources and the role of lug-imprint orientation on erosion is presently under investigation at the Hilton site using recently constructed receivers collecting soils eroded from individual wheelings.

4 DISCUSSION

The total amount of rill erosion and its relative importance in denudation processes increased during intense storms. Such storms are more likely to occur in summer due to the development of convective storm cells (BARRETT 1976). Convective cells may develop because colder, drier air above moist air creates potential instability in the lower troposphere. Instability may occur through movement of polar maritime air over warmer surfaces, due to heating of humid air or line convection along cold fronts (WEBB 1984). Analysis of rainfall records in the West Midlands indicates a marked increase in maximum rainfall intensities between May and September (FULLEN 1984, FULLEN & REED 1986). Reports of erosion on British arable soils have tended to coincide with the spring to late summer period of convective storm activity (MORRIS 1942, CATT et al. 1975, EVANS & NORTCLIFF 1978, FOSTER 1978, REED 1979b, MARTIN & MORGAN 1980). These observations suggest a close relationship between rill development and seasonal rainfall patterns.

Improvements in tractor efficiency, stability and power enable cultivation of steep slopes and arable use of increasingly steep slopes is a current trend in British agriculture (FULLEN 1985a). The marked increase in erosion and rill development particularly evident on runoff plots with slopes $>13°$ questions the wisdom of arable cultivation on slopes up to $16°$, which is common in the study area. These steep slopes are known to be the main sites of severe erosion (REED 1979a,b).

Field studies indicate a close relationship between agricultural operations and subsequent rill development. Tractor wheelings and drill lines can provide efficient channels for runoff. Incision along these lines can result in confined rill erosion and, in extreme cases, confined gully erosion. Erosion may break out of cultivation lines and form dendritic patterns of rills or, where there is a combination of confined and unconfined systems, a trellised pattern. Occasionally, these incise into the B horizon to form unconfined gullies.

Numerous agricultural practices contribute to rill erosion. These include up and down slope cultivation, cultivation of steep slopes, multiple tractor passes and trafficking on moist soils. Continual cultivation of arable soils causes the oxidation and depletion of soil organic matter (VORONEY et al. 1981), and diminished organic content increases the erodibility of soils (DE MEESTER & JUNGERIUS 1978, FULLEN & REED 1986). Hence, continuous cultivation may be contributing to a reduction in the resistance of arable soils to rill erosion.

Throughout northern Europe arable farming is becoming increasingly mechanised, with greater use of increasingly heavy farm machinery, and with greater emphasis on continuous arable cultivation. Recent International Symposia, such as Ghent in 1978 (DE BOODT & GABRIELS 1980), Silsoe in 1980 (MORGAN 1981) and Florence in 1982 (PRENDERGAST 1983) have focussed attention on relationships between intensive arable farming and accelerated soil erosion. The continuation of these trends suggests that agricultural factors will play an increasingly important role in the extent of soil erosion in agricultural northern Europe.

ACKNOWLEDGEMENT

We would like to thank Dr. John Catt and Dr. John Smith for their helpful comments on an earlier draft of this paper, and Ken Muggleston and Andy Morris for their help with field work.

REFERENCES

[AKRAM & KEMPER 1979] AKRAM, M. & KEMPER, W.D.: Infiltration of soils as affected by the pressure and water content at the time of compaction. Soil Science Society of America Journal, **43**, 1080–1086.

[BARRETT 1976] BARRETT, E.C.: Cloud and Thunder. The Climate of the British Isles. T.J. Chandler & S. Gregory (Eds.), 199–210. Longman, London.

[CATT et al. 1975] CATT, J.A., KING, D.W. & WEIR, A.H.: The soils of Woburn Experimental Farm I Great Hill, Road Piece and Butt Close. Rothamsted Experimental Station Report for

1974, Part 2, 5–28. Lawes Agricultural Trust, Harpenden.

[DE BOODT & GABRIELS 1980] DE BOODT, M. & GABRIELS, D. (Eds.): Assessment of Erosion. Wiley, Chichester.

[DE HAAN & VAN DER VALK 1970] DE HAAN, F.A.M. & VAN DER VALK, G.G.M.: Effects of compaction on physical properties of soil and root growth of ornamental bulbs. Proceedings of the 1st International Flower Bulbs Symposium, Noorwijk, The Netherlands, 326–332.

[DE MEESTER & JUNGERIUS 1978] DE MEESTER, T. & JUNGERIUS, P.D.: The relationship between the soil erodibility factor K (Universal Soil Loss Equation), aggregate stability and micromorphological properties of soils in the Hornos area, S. Spain. Earth Surface Processes, **3**, 379–391.

[DE PLOEY 1981] DE PLOEY, J.: Crusting and time-dependent rainwash mechanisms on loamy soil. Soil Conservation Problems and Prospects. R.P.C. Morgan (Ed.), 139–152, Wiley, Chichester.

[DOUGLAS 1970] DOUGLAS, I.: Sediment yields from forested and agricultural lands. The Role of Water in Agriculture. J.A. Taylor (Ed.), University College of Wales Aberystwyth Memorandum No. 12, 57–88, Pergamon Press, London.

[EVANS 1980] EVANS, R.: Mechanics of water erosion and their spatial and temporal controls: an empirical viewpoint. M.J. Kirkby & R.P.C. Morgan (Eds.). Soil Erosion, 109–128, Wiley, Chichester.

[EVANS & NORTCLIFF 1978] EVANS, R. & NORTCLIFF, S.: Soil erosion in north Norfolk. Journal of Agricultural Science, Cambridge, **90**, 185–192.

[FOSTER 1978] FOSTER, S.: An example of gullying on arable land in the Yorkshire Wolds. The Naturalist, **103**, 157–161.

[FULLEN 1984] FULLEN, M.A.: An investigation of rainfall, runoff and erosion on fallow arable soils in east Shropshire. Unpub. Ph.D. thesis, Council for National Academic Awards.

[FULLEN 1985a] FULLEN, M.A.: Erosion of arable soils in Britain. International Journal of Environmental Studies, **26**, 55–69.

[FULLEN 1985b] FULLEN, M.A.: Compaction, hydrological processes and soil erosion on loamy sands in east Shropshire, England. Soil & Tillage Research, **6**, 17–29.

[FULLEN & REED 1986] FULLEN, M.A. & REED, A.H.: Rainfall, runoff and erosion on bare arable soils in east Shropshire, England. Earth Surface Processes and Landforms **11**, 413–425.

[GABRIELS et al. 1977] GABRIELS, D., PAUWELS, J.M. & DE BOODT, M.: A quantitative rill erosion study on a loamy sand in the hilly region of Flanders. Earth Surface Processes, **2**, 257–259.

[GAHEEN & NJØS 1977] GAHEEN, S.A. & NJØS, A.: Long term effects of tractor traffic on infiltration rate in an experiment on a loam soil. Agricultural University of Norway, Department of Soil Fertility Management, Report 87.

[GREENLAND 1977] GREENLAND, D.J.: Soil damage by intensive arable cultivation: temporary or permanent? Philosophical Transactions of the Royal Society, London, **B 281**, 193–208.

[LINDSTROM & VOORHEES 1980] LINDSTROM, M.J. & VOORHEES, W.B.: Planting wheel traffic effects on interrow runoff and infiltration. Soil Science Society of America Journal, **44**, 84–88.

[MARTIN & MORGAN 1980] MARTIN, L. & MORGAN, R.P.C.: Soil erosion in mid-Bedfordshire. Atlas of Drought in Britain 1975–76, J.C. Doornkamp, K.J. Gregory and A.S. Burn (Eds.), 47, Institute of British Geographers, London.

[MORGAN 1981] MORGAN, R.P.C. (Ed.): Soil Conservation Problems and Prospects. Wiley, Chichester.

[MORRIS 1942] MORRIS, F.G.: Severe erosion near Blaydon, County Durham. Geographical Journal, **100**, 257–261.

[MOSLEY 1974] MOSLEY, M.P.: Experimental study of rill erosion. Transactions of the American Society of Agricultural Engineers, **17**, 909–913, 916.

[PRENDERGAST 1983] PRENDERGAST, A.G. (Ed.): Soil Erosion. Abridged Proceedings of the Workshop on 'Soil erosion and conservation: assessment of the problems and the state of the art in E.E.C. Countries'. Florence 19–21 October 1982, Commission of the European Communities, Luxembourg.

[REED 1979a] REED, A.H.: Accelerated erosion of arable soils with special reference to the West Midlands. Unpub. Ph.D. thesis, University of Keele.

[REED 1979b] REED, A.H.: Accelerated erosion of arable soils in the United Kingdom by rainfall and runoff. Outlook on Agriculture, **10**, 41–48.

[REED 1983] REED, A.H.: The erosion risk of compaction. Soil and Water, **11**, 29, 31, 33.

[SOANE et al. 1980] SOANE, B.D., BLACKWELL, P.S., DICKSON, J.W. & PAINTER, D.J.: Compaction by agricultural vehicles: a review. Soil & Tillage Research, **1**, 207–237.

[SOIL SURVEY OF ENGLAND AND WALES 1983] SOIL SURVEY OF ENGLAND AND WALES: 1:250,000 Soil Association Map of England and Wales.

[TROUSE 1966] TROUSE, A.C.: Alteration of the infiltration permeability capacity of tropical soils by vehicular traffic. Proceedings of the 1st Pan-American Soil Conservation Congress, São Paulo, Brazil, 1103–1109.

[VORONEY et al. 1981] VORONEY, R.P., VAN VEEN, J.A. & PAUL, E.A.: Organic C dynamics in grassland soils 2. Model validation and simulation of the long term effects of cultivation and rainfall erosion. Canadian Journal of Soil Science, **61**, 211-224.

[WEBB 1984] WEBB, J.D.C.: Thunder in Britain and the abnormal June of 1982. Weather, **39**, 50–58.

[YOUNG & WIERSMA 1973] YOUNG, R.A. & WIERSMA, J.L.: The role of rainfall impact in soil detachment and transport. Water Resources Research, **9**, 1629–1636.

Address of authors:
Michael A. Fullen & Alan Harrison Reed
Environmental Sciences Section, School of Applied Sciences, The Polytechnic, Wolverhampton WV1 1LY, U.K.

THRESHOLD CONDITIONS FOR INCIPIENT RILLING

D. Torri, M. Sfalanga, Firenze
G. Chisci, Palermo

SUMMARY

Surface runoff reaches its maximum power both as a detaching and transporting agent once it is channelled into rills. Consequently the evaluation of the threshold conditions for rilling is very important in order to forecast and control soil loss.

Some laboratory experiments were carried out, under simulated rain, in order to gather more information on threshold conditions. Four types of soil were tested with their textures ranging from sandy loam to clay.

Critical Froude number and critical shear stress of runoff were evaluated and compared with some soil characteristics. Critical shear stress was found to be correlated to soil shear strength as measured with a vane-test apparatus. The ratio between critical shear stress and soil shear strength shows a small residual variation which can be ascribed to experimental errors. The measured value of the ratio agrees with data collected by other researchers.

ISSN 0722-0723
ISBN 3-923381-07-7
©1987 by CATENA VERLAG,
D–3302 Cremlingen-Destedt, W. Germany
3-923381-07-7/87/5011851/US$ 2.00 + 0.25

1 INTRODUCTION

Rills are defined as 'small intermittent water courses with steep sides, usually a few inches deep and, hence, no obstacle to tillage operations' (S.C.S.A. 1982).

Rills probably represent the main features through which soil is eroded on slopes. In fact surface runoff reaches its maximum power both as detaching and transporting agent once channelled into rills. This concept was stressed by MORGAN (1977) when he found that the amount of sediment removed from rills was about 40 times the amount removed from the interrill areas on an 11° degree sloping sandy soil.

Consequently, evaluation of the threshold conditions for incipient rilling is very important in order to apply conservation measures able to control soil loss effectively.

This paper aims to add more information about critical threshold conditions for incipient rilling with special attention to clay-rich soils.

2 SCIENTIFIC BACKGROUND

Rills are formed when the detaching and entraining forces due to surface runoff overcome the forces keeping the soil particles and aggregates in place.

Some information of rill processes can be found in the literature on channel dynamics and critical conditions of incipient motion of grain for cohesive sediments (VANONI 1977, 107–114, RAUDKIVI 1976, 262–281). Many authors (SUNDBORG 1956, DUNN 1959, SMERDON & BEASLEY 1961, FLAXMAN 1963, PARTHENIADES & PAASWELL 1970) have shown that incipient motion depends on soil characteristics. Many soil parameters (i.e., soil shear strength, plasticity index, clay content, etc) have been related either to the critical power of runoff.

Power functions of the discharge rate were proposed in order to describe detachment and transport in rills. KIRKBY (1978) indicated an exponent of 2 for runoff discharge while QUANSAH (1985) reported an exponent of 1.4. In the meanwhile, ONSTAD & FOSTER (1975) proposed a 2-factor estimator (total runoff volume times the cubic root of the peak runoff rate).

SAVAT (1979) proposed the Froude number (Fr) as a criterion for incipient rilling conditions and found that the critical Froude number increased with increasing median grain size. DE PLOEY (1983) observed a critical Fr-value of about 2–3 for loamy soils.

A combination of critical Froude number and of washload concentration (critical value for rills 20000 ppm) was proposed by BOON & SAVAT (1981) to identify susceptibility to rilling.

From field observations DE PLOEY (1983) found out that rills already appear on 2–3° sloping loamy soils. GOVERS (1985) explained the observed values of critical slope angle for rill formation on the base of critical shear velocity of runoff.

CHISCI et al. (1985), using laboratory data, related soil erosion rate to runoff shear stress at incipient rilling.

Other authors (GRISSINGER & ASMUSSEN 1963, ALTSCHAEFEL 1965) have pointed out that chemical and environmental factors affect threshold conditions and this is confirmed by the recent paper by TARCHITZKY et al. (1984) on the effect of water quality on infiltration.

3 EXPERIMENTAL APPARATUS AND PROCEDURES

The experiments were made in the laboratory. Slope angle, runoff and rain intensity were manipulated in order to define the threshold conditions for rilling.

Rain was simulated using two oscillating nozzles sprinkling downwards (BAZZOFFI et al. 1980). The main characteristics of the simulated rain can be summarized as follows: the median drop diameter was 1.9 mm; the kinetic energy per unit of rain mass was 24.1 Joule/kg. The rain intensities usually simulated equalled 30 and 60 mm/h (intermittence was of 0.8 and 0.2 s respectively) which are typical values of spring and summer storms. Some runs with rain intensity of 15 and 110 mm/h were added in order to enlarge the explored range.

Four types of soil (tab.1) were used in the laboratory tests. They were collected in the Ao layer, air dried and sieved through a 4-mm net. The material passing through the net was settled into containers 200 cm long, 50 cm wide and 10 cm deep. Soil was free to drain from the bottom and from the downslope side of the container.

It was possible to vary the slope of the soil containers from 0.5° to 17°.

Some tests were made adding clear

φ (mm)	SOIL			
	SP Fluventic Dystrochrepts	GB Vertic Eutrochrepts	F1 Typic Udorthent	VI Vertic Xerochrepts
<0.002	11.0	44.0	52.0	52.0
0.002–0.05	14.5	27.5	36.0	44.1
0.05–2.0	74.3	20.0	10.5	3.9
2.0–4.0	0.2	4.5	1.5	0.0
USDA text classif.	sandy loam	clay	clay	silty clay
τ (kPa)	3.5	13.0	9.5	21.0

Table 1: *Some characteristics of the tested soils.*

water from the upslope side of the soil container in order to increase the runoff shear stress.

Velocity and depth of the water flow were calculated using a computer programme proposed by SAVAT (1980) because its predictions fitted the experimental data fairly well. The input data needed in order to run the programme are unit discharge, water temperature, sediment concentration, slope angle and the diameter corresponding to the 90th percentile of the soil textural distribution. Runoff velocity and depth were used in order to calculate the Froude number (Fr) and the runoff shear stress (τ_r):

$$F_r = \frac{v}{\sqrt{gR}} \quad (1)$$

$$\tau_r = \varrho g R \tan\gamma \quad (2)$$

where
v = mean flow velocity;
R = hydraulic radius;
g = acceleration of gravity;
ϱ = fluid density;
γ = slope angle.

It must be noted that both F_r and τ_r are referred to the flow conditions at the lower side of the soil container.

Every test was made after having prewetted the soil surface using rainfall of low intensity. Prewetting lasted till incipient runoff.

The tests lasted for almost 40 min while rain intensity, added runoff (if present) and slope were kept constant.

Immediately after each test soil shear strength was measured using a vane-test apparatus (diameter and height of the vane-cross were equal to 12,4 mm). Soil cohesion was measured after having implanted the vane apparatus into the soil till its upper side was at the soil surface level. Consequently the soil explored by the vane was limited to the first 12.4 mm from the soil surface.

4 RESULTS AND DISCUSSION

The presence or absence of rill incision was determined after each test. A sample was considered to be rilled when one rill (at least 5 cm long, 0.5 cm deep and 1–2 cm wide) was incised into the soil. When the soil surface showed no sign of incision the test was classified as unrilled (plane soil surface at the end of the run). Some intermediate situations, in which small rills, not fulfilling the above mentioned conditions, were inceased were defined as quasi-rilled tests.

Measured erosion rates (average over the last 10 min of each run) are plotted versus runoff shear stress in fig.1. This representation of the data has been chosen after having compared them with the threshold conditions expressed in term of Froude number as proposed by SAVAT (1979).

The Froude number showed some capacity to discriminate rilled from unrilled tests (tab.2). The data for the soil SP were clearly subdivided at $Fr = 0.45 - 0.5$ but the two quasi-rilled tests were mixed with the rilled ones. The quasi-rilled tests of the soil GB and F1, on the contrary, were mixed with the unrilled ones. Moreover two unrilled tests of the soil GB were mixed with the rilled ones.

The critical values of the Froude number are given in tab.2. Those data indicate that the Froude number decreases as the median grain size increases. This contrasts with the observations of SAVAT (1979) and BOON & SAVAT (1981). Consequently the Froude number was abandoned in favour of the runoff shear stress.

Examining fig.1 it appears immediately that erosion rate increases sharply when a certain value characteristic of each soil, is reached and rills begin to form. This agrees with the results described by GOVERS (1985) and GOVERS & RAUWS (1986).

During some runs it was not clear if rilling was incipient or not (fig.1). This implies that the passage from an interrill to a rill situation takes place smoothly, in agreement with the observations made by SAVAT (1979). The smoothness of the passage can be explained considering that runoff forces act selectively on the soil particles till the coarsest grains can be entrained (GOVERS 1985). When the runoff forces are close to the threshold conditions for the entrainment of the coarsest grains rilling is incipient. In this situation, however, no real rill can develop but some evidence of rilling can be

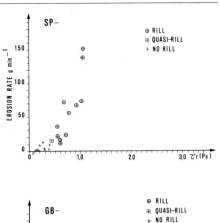
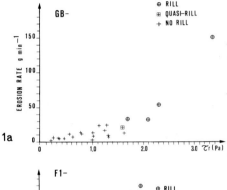
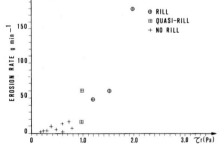
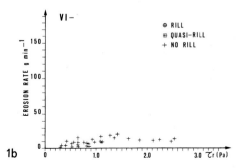

Figure 1: *Relationships between erosion rate, rill formation and runoff shear stress.*

Soil	clay (%)	Percentiles of the text. distr. (μm)			τ_{rcr} (Pa)	Froude Number
		D90	D75	D50		
VI	52	22	8	2	> 2.60	> 3.0
F1	53	70	12	2	0.95–1.1	1.0–1.4
GB	46	950	90	4	1.70–2.0	0.80–0.95
SP	11	310	240	160	0.45–0.50	0.45–0.52
(*)	0–15	45–120	20–110	15–105	0.90–1.22	–
(*) Data from GOVERS (1985)						

Table 2: *Textural distribution characteristics of the tested soils and runoff critical shear stress and Froude number.*

present.

Our data allow only an estimation of the range of τ_r where the critical runoff shear stress (τ_{rcr}) lies. Among the soils studied only soil VI never showed any sign of rilling. Consequently only a value of τ_r lower than τ_{rcr} can be estimated.

The ranges of τ_{rcr} are given in tab.2 together with some data concerning clay content and three percentiles of the textural distribution of the soils. It appears from these data than no clear relationship exists between τ_{rcr} and the given percentiles of the textural distribution. Only an inverse relation seems to exist between the critical shear stress and the D75. However, if the data collected by GOVERS (1985) are also considered this relation no longer holds.

The clay content of the soil does not appear to be well correlated to τ_{rcr}. The good relationship (fig.2) found by SMERDON & BEASLEY (1961) agrees with the data of the soil SP and GB but it does not fit F1 and the data of GOVERS (1985). Also the soil VI appears to scatter upwards.

Comparison between critical runoff shear stress and soil shear strength (fig.3) seems to indicate that the latter is the better related to τ_{rcr} among the studied soil characteristics.

This can be explained (FLAXMAN 1986) as follows: an aggregate or a particle belonging to a cohesive soil is not merely linked to the soil mass only by its weight, but also by other forces. A measure of those forces is soil shear strength (τ_s). Consequently the following critical shear ratio can be proposed:

$$T = \frac{\tau_{rcr}}{\tau_s} \qquad (3)$$

Using data given in tab.1 and tab.2 the experimental critical shear ratio T ranges between 0.0001 and 0.0005. These low values of T can be tentatively explained considering that soil shear strength, as measured with a vane-shear apparatus, is certainly larger than the shear strength linking the most weakly-attached particles or aggregates to the soil mass. Actually, the vane shear strength is mainly due to particles and aggregates buried in the soil while runoff detaches primarily the most exposed in the flow. In the meanwhile, τ_r is an underestimation of the pressures exerted by runoff over a soil particle because local turbulence and Bernoulli's lift are not taken in account. Moreover runoff was considered uniformly spread over the soil surface width in order to calculate runoff depth and τ_r. Actually some micro-concentration of runoff was present making τ_r an underestimation of the maximum shear stress.

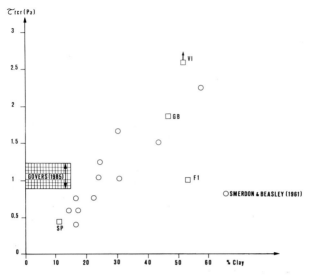

Figure 2: *Critical shear stress in relation to clay content.*

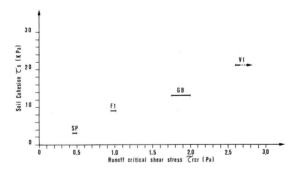

Figure 3: *Soil shear strength versus runoff critical shear stress.*

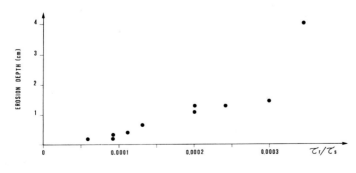

Figure 4: *Relationships between erosion depth and the ratio τ_r/τ_s (after ABDEL-RAHAMANN 1964).*

If the mean values of τ_{rcr} are used for each soil, instead of their interval of variation, the range in which T lies restricts to 0.00011–0.00014. This range seems sufficiently small as to be ascribed to experimental errors. This means that the shear ratio T could be considered as a relatively constant soil parameter characterizing the threshold condition for incipient rilling.

Obviously more data are needed in order to confirm this statement. At present a T-value close to the one here proposed can be derived from the data of GOVERS (1985) if the soil cohesion values given by SAVAT (1979) for similar soils ($\tau_s \simeq 1 - 100 kPa$) are used. Moreover data of DUNN (1959), collected with an experimental procedure very different from ours, are characterized by a T-value ranging between 10^{-1} and 10^{-4}, in partial agreement with our range. Other evidence, supporting both the hypothesis that soil shear strength is strictly linked to rilling and the range of values of the shear ratio T can be found in the data of ABDEL-RAHAMANN (1964) (also quoted in VANONI 1977). His experiments were made using a flume 10.5 m long, with a clay sediment in it. Only runoff was simulated, each run lasting 10 days. At the end of each test soil shear strength was measured using a 3.5 cm-long vane apparatus. Erosion, expressed as surface lowering, was observed to take place during the first 80 hours. Then a thin layer of sticky material was formed at the bed surface, preventing further erosion. This indicate that soil resistance is strongly influenced by soil chemical and mineralogical characteristics and that it can vary while chemical reactions change soil physical characteristics. Nevertheless, the measured soil shear strength can be referred to the soil condition before the sticky layer was formed because the soil explored by the vane was mainly the soil underneath the sticky layer. Consequently a shear ratio can be defined and compared with the total erosion depth. Fig.4 shows that a clear relation exists between these two variables. This relation stresses both the preminent role of τ_s and the fact that the value of critical shear ratio is close to 10^{-4}, provided that soil shear strength is measured with a vane apparatus.

5 CONCLUSION

The data collected in a set of laboratory experiments indicate that there is a smooth passage from a situation of interrill erosion to a situation of rill erosion, conforming the observations made by SAVAT (1979) on the actual presence of an area of uncertainty in the threshold conditions for incipient rilling.

Once the runoff shear stress has overcome a critical value rills are incised into the soil while erosion rate sharply increases. The critical value of runoff shear stress seams strictly linked to the soil shear strength as it was determined with a vane-test apparatus.

The experimental data indicate that the ratio between the runoff critical shear stress and the soil shear strength does not vary widely for the different types of soil tested in the laboratory. Notwithstanding further investigations are needed due to the small amount of available data.

It must be stressed that soil shear strength can vary during a rainstorm because chemistry and mineralogy of soil can interact changing soil resistance of the top layer.

ACKNOWLEDGEMENT

The authors would like to thank Prof. R. Bryan for his helpful assistence in editing and improving this paper.

REFERENCES

[ABDEL-RAHAMANN 1964] ABDEL-RAHAMANN, N.M.: The effect of flowing water on cohesive beds. Contribution n. 56, Versuchsanstalt für Wasserbau und Erdbau an der Eidgenössischen Technischen Hochschule, Zürich, Switzerland, 1–114.

[ALTSCHAEFFL 1965] ALTSCHAEFFL, A.G.: Discussion of 'Erosion and deposition of cohesive soils' by E. Partheniades. Journal of Hydraulics Division, ASCE, **91**, No. HY5, Proc. Paper 4464, Sept. 1965, p. 301.

[BAZZOFFI et al. 1980] BAZZOFFI, P., TORRI, D. & ZANCHI, C.: Stima dell'erodibilità dei suoli mediante simulazione di pioggia in laboratorio. Annali Istituto Sperimentale Studio Difesa del Suolo, **XI**, 129–140.

[BOON & SAVAT 1981] BOON, W. & SAVAT J.: A nomogram for the prediction of rill erosion. In: Soil Conservation: Problems and Prospects, Morgan R.P.C. ed., J. Wiley, Chichester, 303–320.

[CHISCI et al. 1985] CHISCI, G., SFALANGA, M. & TORRI, D.: An experimental model for evaluating soil erosion on a single-rainstorm basis. In: Soil Erosion and Conservation. El Swaify, S.A., Moldenhauer, W.C., Lo, A., eds.: Soil Conservation Society of America, Ankeny, Iowa. 558–565.

[DE PLOEY 1983] DE PLOEY, J.: Runoff and rill generation on sandy and loamy topsoils. Z. Geomorph. N.F., Suppl.-Bd. **46**, 15–23.

[DUNN 1959] DUNN, I.S.: Tractive resistance of cohesive channels. Journal of the Soil Mechanics and Foundations Division, ASCE, No. SM3, Proc. Paper 2062, 1–24.

[FLAXMAN 1963] FLAXMAN, E.M.: Channel stability in undisturbed cohesive soils. Journal of the Hydraulics Division, ASCE, **89**, No. HY2, Proc. Paper 3462, Mar. 1963, 87–96.

[FLAXMAN 1966] FLAXMAN, E.M.: Discussion of 'Sediment transportation mechanics: initiation of motion' by the Task Committee on Preparation of Sediment Manual, Committee on Sedimentation of the Hydraulics Division. V.A. Vanoni, Chmn., Journal of the Hydrasulics Division, ASCE, **96**, No. HY6, Proc. Paper 4959, Nov. 1966, 245–248.

[GOVERS 1985] GOVERS, G.: Selectivity and transport capacity of thin flow in relation to rill erosion. CATENA, **12**, 35–49.

[GOVERS & RAUWS 1986] GOVERS, G. & RAUWS, G.: Transporting capacity of overland flow on plane and on irregular beds. Earth Surface Processes and Landforms, **II**, 505–524.

[GRISSINGER & ASMUSSEN 1963] GRISSINGER, E.H. & ASMUSSEN, L.E.: Discussion of 'Channel stability in undisturbed cohesive soils' by E.M. Flaxman, Journal of the Hydraulics Division, ASCE, **89**, No. HY6, Proc. Paper 3708, Nov. 1963, 259–264.

[KIRKBY 1978] KIRKBY, M.J.: Implication for sediment transport. In: Hillslope Hydrology. M.J. Kirkby ed., Wiley, Chichester, 325–364.

[MORGAN 1977] MORGAN, R.P.C.: Soil erosion in the United Kingdom: field studies in the Silsoe area, 1973–75. Nat. Coll. Agr. Engng. Silsoe, Occasional Paper **4**.

[ONSTAD & FOSTER 1975] ONSTAD, C.A. & FOSTER, G.R.: Erosion modeling on a watershed. Transactions of the ASAE, **18**, 2, 228–292.

[PARTHENIADES & PAASWELL 1970] PARTHENIADES, E. & PAASWELL, R.E.: Erodibility of channels with cohesive boundary. Journal of the Hydraulics, ASCE, **96**, No. HY3, Proc. Paper 7156, March 1970, 775–771.

[QUANSAH 1985] QUANSAH, C.: Rate of soil detachment by overland flow, with and without rain, and its relationship with discharge, slope steepness and soil type. In: Soil erosion and conservation. S.A. El Swaify, W.C. Moldenhauer, A. Lo eds., Soil Cons. Soc. of America, Ankeny, Iowa, 406–423.

[SAVAT 1979] SAVAT, J.: Laboratory experiments on erosion and deposition of loess by laminar sheet flow and turbulent rill flow. In: Seminar on agricultural soil erosion in temperate non mediterranean climate. Strasbourg, 139–143.

[SAVAT 1980] SAVAT, J.: Resistence to flow in rough supercritical sheet flow. Earth Surface Processes, **5**, 103–122.

[S.C.S.A. 1982] S.C.S.A.: Resource conservation glossary. Soil Cons. Soc. of America.

[**SMERDON & BEASLEY 1961**] SMERDON, E.T. & BEASLEY, R.P.: Critical tractive forces in cohesive soils. Agricultural Engineering, Jan. 1961, 26–29.

[**SUNDBORG 1956**] SUNDBORG, A.: The river Klaralven, a study of fluvial processes. Geografiska Annaler, 127–316.

[**TARCHITZKY et al. 1984**] TARCHITZKY, J., BANIN, A., MORIN, J. & CHEN, Y.: Nature, formation and effects of soil crusts formed by water drop impact. Geoderma, **33**, 135–155.

[**VANONI 1977**] VANONI, V.A.: Sedimentation Engineering. ASCE Task Committee, New York, N.Y.

Addresses of authors:
D. Torri
C.N.R. Centro Studio Genesi,
Classificazione e Cartografia Suolo
P. le Cascine 15
50144 Firenze, Italy
M. Sfalanga
M.A.F. Istituto Sperimentale Studio Difesa Suolo
P.za D'Azeglio 30
50100 Firenze, Italy
G. Chisci
Ist. Agronomia Gen. e Coltivazioni Erbacee, Fac. Agraria, Univ. Studi Palermo
Viale delle Scienze
90128 Palermo, Italy

THE INITIATION OF RILLS ON PLANE BEDS OF NON-COHESIVE SEDIMENTS

G. **Rauws**, Leuven

SUMMARY

The initiation of rills in plane non-cohesive beds with a different boundary roughness was studied in laboratory experiments. On slopes $> 2°-3°$ rills originate from a headcut when the shear velocity of the flow exceeds 3.2 to 3.4 cm/s. On these slopes this hydraulic threshold coincides with the appearance of turbulent spots in supercritical flow conditions. The resulting flow structure may be of high complexity. On slopes smaller than 2° the sedimentological response to the flow is totally different.

1 INTRODUCTION

KIRKBY (1980) distinguished two views in modelling the initiation of rills. One view is that rill generation is a balance between infilling by interrill processes and erosion by rill processes. The other view is that rills begin as soon as an hydraulic threshold is exceeded. This hydraulic threshold is a function of the resistance of the bed material. Both views are not contradictory. The definition of a hydraulic threshold is of primary importance. Different authors draw attention to the hydraulic conditions of rill initiation (SAVAT 1976, LAUSHEY & BRAUSCH 1979, SAVAT 1979, KIRKBY 1980, MOSS et al. 1982, SAVAT & DE PLOEY 1982, DE PLOEY 1983, MERRITT 1984, GOVERS 1985). These authors proposed different hydraulic parameters to predict rill formation: unit discharge, stream power, Froude number, critical bottom velocity and shear velocity. Few experimental data are available to verify these proposed thresholds in a wide range of experimental conditions. Despite the interesting analyses by MOSS et al. (1980) and SAVAT & DE PLOEY (1982) it is still impossible to explain a hydraulic threshold from fluid dynamic considerations. The insight in the fluid dynamics of thin water films and their impact on sediment transport and erosion is very limited.

This paper reports the results of an introductory study on the formation of small channels (rills) in a plane bed of loose sediments. Three sets of experiments are described. The first experiments were conducted to define a relevant hydraulic parameter for rill initiation. The relation between the flow patterns of thin water films and the hydraulic flow parameters is examined in a second set. At the end some obser-

ISSN 0722-0723
ISBN 3-923381-07-7
©1987 by CATENA VERLAG,
D–3302 Cremlingen-Destedt, W. Germany
3-923381-07-7/87/5011851/US$ 2.00 + 0.25

vations are reported on the sedimentary structures formed as a result of different flow patterns and different hydraulic conditions.

2 LABORATORY EXPERIMENTAL SET UP

Laboratory experiments were conducted on a 2.0 m long and 0.4 m wide well drained tray with an adjustable slope. The rate of water flow was controlled by a pressure device. No rain was applied. Slope of the tray varied from 1.5° to 8°. In a first series of experiments the tray with a 0.1 m thick layer of almost cohesionless silica flour, with a median grain size of 90 μm. Afterwards the experimental runs were repeated on a silica flour bed with a median grain size of 28 μm. Grain size distribution of this material closely resembles eolian loess. Surface velocity of the flow was measured using aluminium dust as a tracer. The runs were recorded by video.

3 RESULTS

Rills formed spontaneously in the initially plane bed or the incision of a rill was provoked by successive increases of the flow discharge every 3 minutes until a headcut was formed.

In the first case the different stages in the formation of rills were identifiable. They are described in great detail by MOSS et al. (1982) and MERRITT (1984). A uniform sheet flow developed subsequently preferential stream lines, protochannels (micro-rills prior to headcutting, phot.1) and rills (after headcutting, phot.2). Protochannels are defined as downstream-extending, straight, bilaterally symmetrical, leveed trenches (MOSS et al. 1982). The time required for rill formation is dependant on bed slope and applied discharge. For these experiments it varied between 0.5 and 35 minutes.

In the second case the different stages were not observed as rills initiated immediately from a local headcut. The generation of a headcut is an important stage in the development of rills on plane beds. It coincides with a rapid increase in the transport of sediment load and with a widening and deepening of the shallow protochannel.

A preliminary calculation of the flow parameters using the program of SAVAT (1980) pointed out that the mean values of the flow parameters only provide a rough estimate of the hydraulic threshold. This is due to the spatial variation of the flow pattern. A more accurate determination of the hydraulic threshold requires the values of the flow parameters more or less at the point where the headcutting starts. Accurate calculation of the surface velocity is possible by measuring on the video play-back monitor the advance of the tracer front in a very short time interval (1/5th second). From the value of the surface flow velocity the other flow parameters were computed. For laminar flow over a perfectly smooth surface the following equations are valid (HORTON et al. 1934):

$$u_s = \frac{3}{2}(\frac{gSq^2}{3\nu})^{1/3} \qquad (1)$$

and

$$\bar{u} = \frac{2}{3}u_s \qquad (2)$$

Initiation of Rills

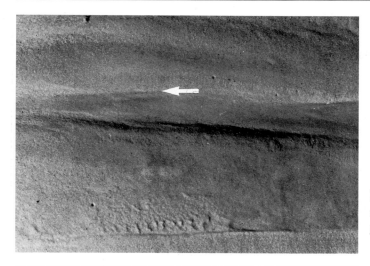

Photo 1: *Protochannel developed on a plane bed.*

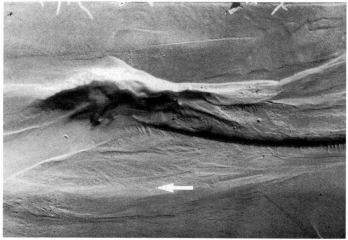

Photo 2: *Headcut developed in a protochannel.*

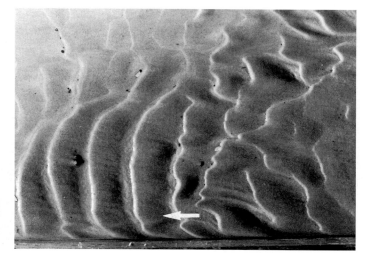

Photo 3: *Ripples developed under turbulent supercritical flow on low slopes.*

where
q = unit discharge of the flow
u_s = surface velocity of the flow
$\bar{u}$ = mean velocity of the flow
S = slope
v = kinematic viscosity of the fluid
g = gravitational acceleration

The following semi-logaritmic equations are valid for turbulent flow. For rough turbulent flow:

$$u_s = 2.5(gRS)^{1/2} ln(30.1 \frac{R}{k_s}) \qquad (3)$$

where
R = hydraulic radius of the flow
k_s = equivalent grain roughness

from which R can be determined by iteration.

For smooth turbulent flow:

$$u_s = 2.5(gRS)^{1/2} ln(9.0(gRS)^{1/2} \frac{R}{v}) \qquad (4)$$

from which R can be determined by iteration. The mean velocity is then computed from:

$$\bar{u} = u_s - 2.5(gRS)^{1/2} \qquad (5)$$

However it is uncertain to what extent the formulae (1) to (5) approximate the real values of the flow parameters in thin water films with a rough boundary. In the case of laminar flow the theoretical surface velocity u_s deviates from its measured value u'_s. This deviation seems to be a function of the relative submergence R/k_s of the roughness elements (PHELPS 1975a). For three sets of data (EMMET 1970, PHELPS 1975a and NITTIM 1977) the measured surface velocity was compared with its computed value, according to (1) and (3). Fig.1 shows that if R/k_s is larger than 4 to 5, the deviation of the observed surface velocity never exceeds 10%. The use of the formulae (1) to (5) only seems to be inappropriate at low values of relative submergence. This ascertainment confirms the delimitation of "large-scale roughness elements" at $R/k_s = 4$ by BATHURST et al. (1981). The flow resistance of large-scale roughness elements is mainly related to the form drag of the elements and of their disposition on the bed.

Fig.2 shows clearly that on plane beds with loose sediments headcutting always occurs on slopes $>3°$ when the shear velocity $u_* = (gRS)^{1/2}$ exceeds 3.2 to 3.5 cm/s. When slope is 2°–3° rilling is possible when $u_* > 3.2$ to 3.5 cm/s. On slopes $< 2°$ it is almost impossible that rills occur, as stated by SAVAT & DE PLOEY (1982) from a literature review of field data. In fact, when slope is lower than 2° very high unit discharges are required to reach a shear velocity of 3.2 cm/s and runoff will rather occur as a sheetflood, without concentration of the flow and without clear incision. For the experiments with the finer silicaflour the threshold seems to be a little higher. This may be due to the somewhat higher cohesion of the fine material (SAVAT & DE PLOEY 1982). The value of the hydraulic threshold is independant of the way of rill formation. There is no difference between the rills formed spontaneously out of protochannels and those provoked by successive increases of the discharge.

4 DISCUSSION

The experiments confirm the hypothesis of GOVERS (1985), who proposed a value of the shear velocity between 3.0 and 3.5 cm/s as a possible threshold for the initiation of rills. The author gave two arguments for this proposal: it was observed that when the shear velocity exceeds 3.2 cm/s the transporting capacity

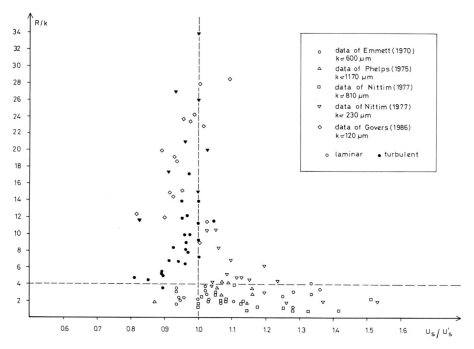

Figure 1: *Ratio of theoretical surface velocity of thin water films to its measured value in function of the relative submergence of the bed roughness elements.*

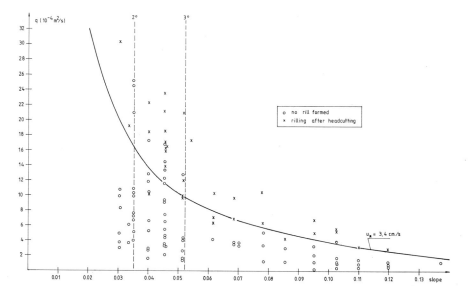

Figure 2: *The initiation of rills in plane beds of loose sediments in function of slope of the bed and unit discharge of the flow.*

of thin liquid films increases rapidly and that beyond the same threshold sediment transport of the flow is no longer size-selective. These observations thus confirm the statement of SAVAT (1979) that rills can only occur in plane beds when the coarser grains move as easily as the finer fractions.

The extrapolation of this threshold value to natural slope conditions seems to be rather difficult at first sight. Until now no experiments with rainflow have been conducted. The coincidence of the increase in transporting capacity of overland flow and the initiation of rills at a shear velocity of 3–3.5 cm/s suggests that this threshold value is valid if the transporting capacity of the flow is limiting factor during rill formation. This is the case whenever there is an important sediment detachment during rill formation. The plane bed surface and the non-cohesive character of the bed material are other limitations. Resembling conditions are encountered if e.g. a dried surface seal on a loamy topsoil is rewetted at once. Surface seal material is then very sensitive to dispersion (BEN-HUR et al. 1985) and to liquefaction (DE PLOEY 1985). Field observations of rill initiation after a long dry period on sealed agricultural fields with loamy topsoils in the region around Huldenberg (Belgium) revealed a complete resemblance to rill generation in loose sediments in a laboratory tray.

However there are indications that the shear velocity threshold is significant in many other situations. SAVAT (1977) found that the influence of raindrop impact on the hydraulics of thin liquid films was rather limited. GOVERS (1985) shows that sediment transport during rainflow stays size-selective when the shear velocity is inferior to 3–3.5 cm/s. Raindrop impact has a definite influence on the transporting capacity of overland flow, but does not change the shear velocity threshold. It seems unlikely therefore that raindrop impact influences in a major way the critical conditions for rill initiation.

GOVERS & RAUWS (1986) found that the transporting capacity of overland flow on irregular beds is dependant on the grain shear velocity (i.e. that part of the total shear velocity that is exerted on the soil particles) of the flow. Laboratory data from experimental runs on irregular beds revealed a sharp increase of the transporting capacity at a grain shear velocity of 3 cm/s. This is an indication that also on irregular slopes the proposed threshold can be relevant. It is clear that more experiments are needed to verify this assumptions in a wide range of situations. This falls however beyond the scope of this paper.

5 FLUID DYNAMIC CONSIDERATIONS

Because the impact of this hydraulic threshold seems to be quite fundamental it can be expected that if the shear velocity exceeds 3.2 cm/s, the fluid pattern changes significantly. Flow of a thin liquid film is highly complex. This is due to the large thickness of the boundary layer relative to the flow depth and to the three-dimensional character of the flow.

5.1 LABORATORY SET-UP

The fluid pattern in thin water films was examined during experiments in a small glass flume of 0.06 m width and 2.0 m length. Fixed beds with different boundary roughness (varying from silt size to median sand) were used. Discharge and slope were adjustable. Flow pattern were identified using potassium permanganate as a dye. Flow hydraulic parameters were calculated with the program of SAVAT (1980).

5.2 OBSERVATIONS

By increasing the unit discharge laminar flow becomes transitional. The onset of transition is characterized by the appearance of small vortices. At sill higher unit discharges turbulent spots appear. These are isolated patches of turbulence (fig.3), occuring in the last phase of the laminar-turbulent transition. The main characteristics of turbulent spots are described in paragraph 5.3. Fig.4 shows the values of the shear velocity at which turbulent spots appear on different slopes. On slopes larger than 2°, turbulent spots appear at a shear velocity of 3.0 to 3.4 cm/s. On these slopes they are often associated with the appearance of roll wave trains. On lower slopes turbulent spots are observed at smaller values of the shear velocity.

This observation suggests that on plane beds of loose sediments rills cannot originate in a true laminar flow. The generation of rills might therefore be correlated with a certain degree of turbulence of the flow or with the presence of certain turbulent structures in the flow.

Flow structure was also identified using a dye during the different stages in the formation of rills in the large tray. Slopes were obviously steeper than 2°. A uniform sheet flow is never observed, as preferential stream lines develop immediately. Flow in the stream lines is always laminar. Flow in protochannels becomes transitional and the developing turbulent spots bring bed material into sus-

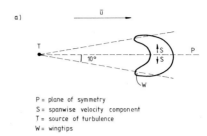

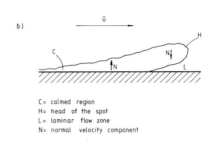

P = plane of symmetry
S = spanwise velocity component
T = source of turbulence
W = wingtips

C = calmed region
H = head of the spot
L = laminar flow zone
N = normal velocity component

Figure 3: *Schematic diagram of a turbulent spot.*
a. Plan view.
b. Lateral view.

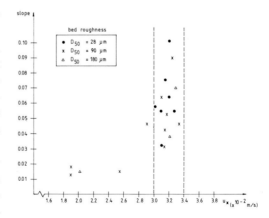

Figure 4: *The appearance of turbulent spots in thin water films in function of slope of the bed and shear velocity of the flow. Each symbol denotes the appearance of a turbulent spot.*

pension. When the protochannels grow in length, the period between the generation of subsequent turbulent spots becomes smaller until at certain points the flow in the protochannels is turbulent most of the time. In that stage a headcut occurs. In fully developed turbulent flow conditions, far beyond the shear velocity threshold, deep and wide channels are dug, but incision deepens gradually without a headcut being formed.

It is concluded that the phenomenon of headcutting by afterflow in non-layered loose sediments of silt and fine sand size occurs in a limited range of flow conditions. Slope must be at least 2° and flow may not be truly laminar or fully turbulent. Therefore headcutting seems to be related to the presence of transition flow structures.

5.3 DISCUSSION

Turbulent spots form spontaneously and randomly in time and space on a smooth bed, but can also be induced by introducing disturbances (EMMONS 1951). Each spot grows in size and spreads laterally as it is convected downstream. The transverse velocity component in a turbulent spot is everywhere directed outwards (i.e. away from the plane of symmetry, fig.3) (WYGNANSKI et al. 1976). The growth of spots and their progressive agglomeration cause an increasing part of the flow to become turbulent. The passage of a spot in a turbulent boundary layer is characterized by a substantial velocity excess near its center line in the vicinity of the flume wall and consequently by a higher wall shear (ZILBERMAN et al. 1977).

Very little is known about sediment entrainment by a turbulent spot, although it is the subject of an extensive research program (BROWAND et al. 1982). Flow structures appearing in transitional flows are thought to be responsible for the generation of some natural bed forms. KARCZ (1970) linked the occurence of erosional flute marks on mud beds with the hairpin vortices and turbulent spots of late transition.

In literature the appearance of turbulent spots is hardly correlated with the hydraulic parameters of the flow. ELDER (1960) states that turbulent spots are generated at all points at which the turbulent velocity fluctuation exceeds a critical intensity, regardless of how disturbances are generated and independant of the Reynolds number. This agrees well with the identification of a critical shear velocity, because this parameter is correlated with the apparent shear stress due to velocity fluctuations especially in the absence of a laminar boundary layer.

Nevertheless transition in thin liquid films on relative steep slopes may differ from "normal" boundary layer transition. AZUMA & HOSHINO (1984) observed in radial flow that any disturbance in the vicinity of the wall, appears as an undulation (wave) of the liquid surface, due to the extremely small thickness of the liquid film. Moreover the flow surface becomes unstable in the subcritical laminar regime, at Froude numbers as low as 0.5 (ISHIHARA et al. 1960, KARCZ & KERSEY 1980). When rills are incised in a plane bed, flow is always supercritical and extensive flow separation with large downstream eddies may occur (SAVAT & DE PLOEY 1982). Photographs of PHELPS (1975a) clearly show that not only the Froude number but also the Reynolds number based on the shear velocity is related to flow separation in laminar flows. The association of transition to turbulence and the appearance of roll wave trains was already mentioned by EMMONS (1951) and PHELPS (1975b).

KARCZ & KERSEY (1980) suggest that transition underneath waves and surges may be similar to transition in a boundary layer under flow with superimposed periodic fluctuation, as studied by KOBASHI & HAYAKAWA (1978). In such flows turbulent patches—closely resembling turbulent spots—appear suddenly in a wave packet, grow rapidly downstream, until the fore part of the patch tilts and encroaches the boundary. The resulting curling configuration was attributed by KOBASHI & HAYAKAWA (1978) to the presence of secondary flow (vertical and transverse flow components), commonly invoked in the interpretation of local scour. Only MOSS et al. (1982) have explicitly studied the relation between flow structure and the formation of small channels, and have suggested that secondary flow is responsible for the initiation of protochannels and channels. The observation of roll waves and turbulent spots (with outwards directed transverse and downwards directed normal velocity components as inherent characteristics) as well as the analyses of fluid patterns in thin water films by KARCZ & KERSEY (1980) and by SAVAT & DE PLOEY (1982) imply that a more complex system is involved.

5.4 INFLUENCE OF SLOPE ON SEDIMENTARY STRUCTURES

An experimental run was conducted in the 2.0 on 0.4 m tray on a 0.8° slope. The tray was filled with the 90 μm silica flour. Observations were made on the flow patterns and the sedimentary structures formed as a result of successive increases of the unit discharge. The results are shown in tab.1. Protochannels formed in laminar subcritical flow. In the protochannels flow became supercritical with standing waves. When the shear velocity was about 2.0 cm/s turbulent spots brought material in suspension. No headcut was formed. When the discharge was increased, flow outside the protochannel became transitional and a rhomboid bed configuration was formed. At still higher flow discharge, in fully turbulent conditions, protochannels disappeared as the whole bed was unstable and ripples were formed (phot.3).

From all these observations it is clear that slope is a determining factor concerning the sedimentary structures formed under thin water films. Possible explanations are the complex relationship between critical transition Reynolds number and slope (DE BOER 1985) and—for equal Reynolds numbers—the strong dependance of the Froude number on the slope (SAVAT & DE PLOEY 1982).

The sedimentary structures formed on plane beds of loose sediments on different slopes are compared in tab.2. Data of KARCZ & KERSEY (1980) from experiments on slopes smaller than 1.15° are compared with data of MERRITT (1984) from experiments on a 5° slope. For the data of KARCZ & KERSEY the program of SAVAT (1980) was used to compute the missing flow parame-

S = 0.14 Silica flour D_{50} = 90 μm				
Re	Fr	Surface stability	Flow pattern	Sedimentary structures
90	0.55	no waves	laminar	smooth bed
135	0.68	local standing waves	laminar	protochannel
340	1.08	standing waves	laminar	protochannel
540	1.36	standing waves roll waves	transitional	protochannel and rhomboids
890	1.57		transitional turbulent	rhomboids
2760	1.83		turbulent	
Re = Reynolds number ($Re = q/v$)				
Fr = Froude number ($Fr = \bar{u}/(gR)^{1/2}$)				

Table 1: *Sedimentary structures formed on a plane bed of loose sediments on a 0.8° slope.*

KARCZ & KERSEY (1980)				MERRITT (1984)			
slope ≤ 0.02 well sorted fluvial sand 177–250 μm				slope = 0.087 sandy loam D_{50} = 300 μm			
$u_*^{(1)}$ (cm/s)	turbulence	surface stability	sedimentary structure	$u_*^{(1)}$ (cm/s)	turbulence	surface stability	sedimentary structure
1.2	laminar	subcritical	smooth bed	1.3	laminar	supercritical	smooth bed
1.5	laminar	subcritical	ridges				
1.6	laminar	supercritical	ridges and rhomboids				
1.7	transitional	supercritical	rhomboids				
				2.1	laminar	supercritical	flowlines
				2.8	transitional	supercritical	micro-rills prior to headcutting
$3.1^{(2)}$	turbulent	supercritical	ripples	3.1	transitonal	supercritical	micro-rills after headcutting
(1) mean value							
(2) datum of experiments by the author, value of * as an indication, not to be generalized slope = 0.014 and 0.0175; sand D_{50} = 180 μm							

Table 2: *Sedimentary structures formed on plane beds of loose sediments on different slopes.*

ters. Different bed forms developed in apparently similar experimental conditions with flow discharge of similar magnitude and analogous fluid dynamic conditions. It seems unlikely therefore that the rapidly formed bed configurations described by KARCZ & KERSEY, provide a mechanism for rill generation, as they suggest. To reach a value of the shear velocity of 3.2 cm/s at a slope of 0.02 the water film has to be 0.52 cm thick. In such water films it may be impossible to achieve a linear concentration of the shear stress, necessary for the initiation of rills on a plane bed.

6 CONCLUSION

Experimental laboratory work confirms that the shear velocity of the flow is significant for rill initiation. On plane almost non-cohesive beds rills originate from a headcut when the shear velocity of the flow exceeds 3.0–3.5 cm/s and slope is larger than 2°. From flow analysis it is clear that both laminar-turbulent transition and flow surface instability explain this threshold. Nevertheless the fluid dynamics of thin water films and their role in sediment transport and erosion are largely unexplored. The experiments described show that grain-size itself does not affect the critical threshold of rill initiation in a major way. From field observations and earlier laboratory experiments by SAVAT (1979) it seems that soil physical parameters are much more important. Therefore the initiation of rills on cohesive irregular beds of soil material is currently under study.

ACKNOWLEDGEMENT

Many thanks are expressed to Prof. Dr. J. de Ploey for the constructive comments he made during the experimental work and the preparation of this text.

REFERENCES

[AZUMA & HOSHINO 1984] AZUMA, T. & HOSHINO, T.: The radial flow of a thin liquid film — 1st report, laminar-turbulent transition. Bulletin of JSME **27**, 2739–2746.

[BATHURST et al. 1981] BATHURST, J.C., LI, R.M. & SIMONS, D.B.: Resistance equation for large-scale roughness. Journal of the Hydraulics Division, Proceedings of the ASCE **107**, 1593–1613.

[BEN-HUR et al. 1985] BEN-HUR, M., SHAINBERG, I., KEREN, E. & GAL, M.: Effect of water quality and drying on soil crust properties. Soil Science Society of America Journal **49**, 191–196.

[BROWAND et al. 1982] BROWAND, F.K., OSTER, D., PLOCHER, D. & MC LAUGHLIN, D.: Sediment entrainmant by a turbulent spot. Euromech 156. Mechanics of Sediment Transport, Istanbul, July 1982. Balkema, Rotterdam, 27–32.

[DE BOER 1985] DE BOER, D.H.: A kinematic wave model of overland flow on badland slopes. Ph.D. thesis, Subfac. Fysische Geografie en Bodemkunde, Universiteit van Amsterdam, 91 p.

[DE PLOEY 1983] DE PLOEY, J.: Runoff and rill generation on sandy and loamy topsoils. Zeitschrift für Geomorphologie, Supplement Band **46**, 15–23.

[DE PLOEY 1985] DE PLOEY, J.: Experimental data on runoff generation. Soil erosion and conservation. El-Swaify, S.A., Moldenhauer, W.W., Lo, A. (eds.). Soil Conservation Society of America, Ankeny, 528–539.

[ELDER 1960] ELDER, J.W.: An experimental investigation of turbulent spots and breakdown to turbulence. Journal of Fluid Mechanics **9**, 235–246.

[EMMETT 1970] EMMETT, W.W.: The hydraulics of overland flow on hillslopes. Geological Survey Professional Paper **662-A**, 68 p.

[EMMONS 1951] EMMONS, H.W.: The laminar-turbulent transition in a boundary layer. Journal of the Aeronautical Sciences **18**, 490–498.

[GOVERS 1985] GOVERS, G.: Selectivity and transport capacity of thin flows in relation to rill erosion. CATENA **12**, 35–49.

[GOVERS & RAUWS 1986] GOVERS, G. & RAUWS, G.: Transporting capacity of overland flow on plane and irregular beds. Earth Surface Processes and Landforms, **11**, 505–524.

[HORTON et al. 1934] HORTON, R.E., LEACH, H.R. & VAN VLIET, V.: Laminar sheet-flow. Transactions of the American Geophysical Union **2**, 393–404.

[ISHIHARA et al. 1960] ISHIHARA, T., IWAGAKI, Y. & IWASA, Y.: Discussion on "Roll waves and slug flows in inclined open channels" by P.G. Meyer. Journal of the Hydraulics Division, Proceedings of the ASCE **86**, 45–60.

[KARCZ 1970] KARCZ, I.: Possible significance of transition flow patterns in interpretation of origin of some natural bed forms. Journal of Geophysical Research **75**, 2869–2873.

[KARCZ & KERSEY 1980] KARCZ, I. & KERSEY, D.: Experimental study of free-surface instability and bedforms in shallow flows. Sedimentary Geology **27**, 263–300.

[KIRKBY 1980] KIRKBY, M.J.: Modelling soil erosion processes. Soil erosion. Kirkby, M.J., Morgan, R.P.C. (eds.). Wiley, Chichester, 183–216.

[KOBASHI & HAYAKAWA 1978] KOBASHI, Y. & HAYAKAWA, M.: Development of turbulence through non-steady boundary layer. Structure and mechanisms of Turbulence I, Proceedings Symposium on turbulence, Berlin, August 1977. Fiedler, H. (ed.). Lecture notes in physics **75**, Springer, Berlin, 277–288.

[LAUSHEY & BRAUSCH 1979] LAUSHEY, L.M. & BRAUSCH, L.M.: The geometrics of rill formation on hillsides. International Association for Hydraulic Research, Congress Proceedings **18**, 39–47.

[MERRITT 1984] MERRITT, E.: The identification of four stages during micro-rill development. Earth Surface Processes and Landforms **9**, 493–497.

[MOSS et al. 1982] MOSS, A.J., GREEN, P. & HUTKA, J.: Small channels - their experimental formation, nature and significance. Earth Surface Processes and Landforms **7**, 401–415.

[NITTIM 1977] NITTIM, R.: Overland flow on impervious surfaces. Water Research Laboratory Report **151**, University of New South Wales, Manly Vale, 200 p.

[PHELPS 1975a] PHELPS, H.O.: Shallow laminar flows over rough granular surfaces. Journal of the Hydraulics Division, Proceedings of the ASCE **101**, 367–384.

[PHELPS 1975b] PHELPS, H.O.: Friction coefficients for laminar sheet flow over rough surfaces. Proceedings Institution of Civil Engineers, Part **2 59**, 21–41.

[SAVAT 1976] SAVAT, J.: Discharge, velocities and total erosion of calcareous loess: a comparison between pluvial and terminal runoff. Revue de Géomorphologie Dynamique **24**, 113–122.

[SAVAT 1977] SAVAT, J.: The hydraulics of sheet flow on a smooth surface and the effect of simulated rainfall. Earth Surface Processes **2**, 125–140.

[SAVAT 1979] SAVAT, J.: Laboratory experiments on erosion and deposition of loess by laminar sheet flow and turbulent flow. Proceedings Seminar on agricultural soil erosion in temperate non-mediterranean climate, Strasbourg-Colmar, September 1978. Vogt, H., Vogt, T. (eds.). Université Louis Pasteur, Strasbourg, 139–143.

[SAVAT 1980] SAVAT, J.: Resistance to flow in rough supercritical sheet flow. Earth Surface Processes **5**, 103–122.

[SAVAT & DE PLOEY 1982] SAVAT, J. & DE PLOEY, J.: Sheetwash and rill development by surface flow. Badland geomorphology and piping. Bryan, R., Yair, A. (eds.), Geo Books, Cambridge, 113–126.

[WYGNANSKI et al. 1976] WYGNANSI, I., SOKOLOV, M. & FRIEDMAN, D.: On a turbulent "spot" in a laminar boundary layer. Journal of Fluid Mechanics **78**, 785–819.

[ZILBERMAN et al. 1977] ZILBERMAN, M., WYGNANSKI, I. & KAPLAN, R.E.: Transitional boundary layer spot in a fully turbulent environment. The Physics of Fluids **20**, S258–S271.

Address of author:
Gerrit Rauws
Research Assistant I.W.O.N.L.
Laboratory of Experimental Geomorphology
Redingenstraat 16bis
3000 Leuven, Belgium

A REVIEW OF DISSOLUTIONAL RILLS IN LIMESTONE AND OTHER SOLUBLE ROCKS

D.C. **Ford** and J. **Lundberg**, Hamilton

SUMMARY

Solutional rills are classified and discussed by size, form and supposed mode of origin. The emphasis is on hydraulics of flow, the basic distinction being between channels parallel to the direction of flow (the most common) and those which are transverse to flow (transverse ripples and fluted scallops). The first group are further divided: (1) on very fine-grained rocks MICRORILLS are produced, their form controlled by capillary forces, (2) GRAVITOMORPHIC RILLS are subdivided into (a) those which head at the crest and extinguish downslope (RILLENKARREN); (b) the HORTONIAN TYPE which head below a belt of no channelled erosion and enlarge downslope (RINNENKARREN develop on bare rock, RUNDKARREN develop under a cover); (c) DECANTATION forms which reduce in width and/or depth downslope. Here the solvent is released from an upslope store that may be a point source that produces WANDKARREN, or a diffuse source that produces a suite of parallel DECANTATION FLUTINGS. In nature few rills have had a simple history.

ISSN 0722-0723
ISBN 3-923381-07-7
©1987 by CATENA VERLAG,
D–3302 Cremlingen-Destedt, W. Germany
3-923381-07-7/87/5011851/US$ 2.00 + 0.25

Many must be classified as POLYGENETIC FORMS.

1 INTRODUCTION

This paper reviews the nature and genesis of rill forms developed on soluble rocks. In all of the different forms to be considered experimental or field evidence strongly implies that aqueous dissolution of rock is the only significant erosion process. Within the family of solutional (karst) landforms these are all small-scale to micro-scale (FORD 1980). The general term for such solutional features, "karren", covers both linear forms (fracture controlled and rilling phenomena) and the more circular pit and pan forms, plus many others that are intermediate. The principal discussions and classifications of karren types appear in BOGLI (1960, 1980), SWEETING (1972), PERNA & SAURO (1978), and JENNINGS (1985).

All of the forms to be discussed are known to occur on limestones and marbles. Typically, only the bigger forms will be found on dolomites. Many of them occur on more soluble gypsum and salt, but outcrops of these rocks are comparatively few and small and so have been little studied. In special circumstances some of these rilling phenomena are found on homogeneous, medium to fine grained igneous or volcanic rocks

A.	**CHANNELS PARALLEL TO THE DIRECTION OF FLUID FLOW**	
	1. **Microrills (as in "Rillensteine");** flow is controlled by capillary forces and/or gravity and/or wind; rill width is ~ 1 mm.	
	2. **Gravitomorphic rills**	
	(a)	Heading at the crest, rainfall generated, extinguishing downslope, no decantation: RILLENKARREN (BOGLI 1960 - Type IAa).
	(b)	Hortonian type, heading below a belt of no channelled erosion: RINNENKARREN (sharp rimmed, supposedly developed on bare rock, BOGLI 1960 - Type IAb). RUNDKARREN (rounded rimmed, developed under a soil/vegetation cover, BOGLI 1960 - Type III).
	(c)	Decentation forms, solvent is released from upslope store: WANDKARREN (solvent supplied from point source, many varieties and scales, includes BOGLI 1960 - Type IAb). FLUTINGS (solvent supplied from diffuse source).
	3. **Polygenetic forms:** Mixtures of the gravitomorphic types plus intruding (subsequent) HOHLKARREN (BOGLI 1960 - Type IIB) and solutional WELLS. Generation of residual PINNACLES.	
B.	**CHANNELS TRANSVERSE TO FLUID FLOW** FLUTED SCALLOPS	

Table 1: *Classification of dissolutional rills in soluble rocks.*

such as petitgranite, granodiorite and dense basalts.

Our classification of rilling phenomena is presented in tab.1. The forms are differentiated by a combination of their scale, cross profile, long profile and supposed mode of genesis. The upper width limit of approximately 50 cm roughly divides purely solutional rills from regular stream channels in limestone that carry suspended load and bedload and thus suffer abrasion, etc. Rills are elongated features; the minimum length is of the order of five times the channel width.

In nature there are many transitional or mixed forms. Thus, to some extent, the categorisation in Classes 2b, 2c and 3 must be forced. Reference to these features in the literature is not always precise. We discuss the more common sources of confusion under the relevant sections below.

2 MICRORILLS

The term "Rillensteine" means (micro-) "rilled rock". These microrills are the smallest of the rilling phenomena, typically about one millimetre wide, round bottomed and packed together with characteristic tightly simuous to anastomosing patterns on gentle slopes, becoming straighter and more parallel with increasing slope inclination. Cross sectional form and dimension tend to be constant along the rill. Most reports are of microrills on limestone or gypsum, and they have been particularly noted in hot or arid areas. However, we have seen extensive development in the supralittoral zone on very fine grained limestones of northern Vancouver Island, a cool temperate rain forest climate (phot.1) and in a similar situation in the Gower Peninsula, Wales.

LAUDERMILK & WOODFORD (1932) distinguished four types of microrills (fig.1). They report that widths may reach 3–4 mm but in our experience 0.5–1 mm is predominant. Only one of

Photo 1: *Supralittoral Rillensteine from Vancouver Island, Canada.*

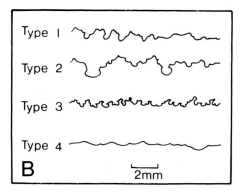

Figure 1: *Types of microrills, after LAUDERMILK & WOODFORD (1932)*
Type 1 tend towards parallel development, are moderately sinuous and rather shallow, Type 2 are more clearly arranged in parallel lines, more sinuous and deeper than Type 1, Type 3 are poorly integrated, shallow, narrow and rough with no single dominant direction. The walls of these three types may be undercut. Type 4 are less intricate, wide and shallow rills with smooth crests and a polished appearance.

these types has been reported by other workers. Microrills on massive, pure but fossiliferous limestones of the Phillipines (LONGMAN & BROWNLEE 1980) and of Chillagoe, Australia (JENNINGS 1981, DUNKERLEY 1983) resemble type 4, having sinuous courses on gentle slopes, becoming straighter with increasing slope angle. In Chillagoe they occur on the sides of solutional rainpits and flutes (see below) in localised patches. On Vancouver Island and in the Gower Peninsula the supralittoral microrills also resemble type 4. In this narrow zone between the sea and the soil, bedrock has been free of soil/vegetation cover for at least several centuries. It is of interest to note that in the inland areas of Vancouver Island bedrock of the same lithology has been exposed to subaerial weathering by deforestation and soil erosion only in the last 50 years. These surfaces show etching patterns similar to types 1 and 3, while neither etching nor microrills appear on surfaces which are currently covered by soil and/or vegetation.

LAUDERMILK & WOODFORD (1932) suggest that on blocks on or in clayey desert soils and in samples from arctic Spitzbergen the rills are probably etched by rising soil solutions; rills may also be caused by dew. Microrills noted by Lundberg on both upper and under surfaces of blocks in the Queensland desert of Australia suggest that both mechanisms may operate concurrently. TRUDGILL (1985) shows typical microrills being formed from organic acid-rich stem flow waters in Yorkshire. However, etching by organic acids can only be of local importance, e.g. it cannot operate evenly over the wide expanses of rilled rock of the supralittoral as in phot.1.

LAUDERMILK & WOODFORD (1932) suggest that scarcity of water

seems to be an essential condition; that the source of water must be occasional desert storms and that on many of the larger rills wind controls the movement of water films. On Vancouver Island >2500 mm per year of rainfall and fine sea spray provide the solvents; there is no scarcity of water. An important point here is that the rocks are very fine-grained and in the Gower Peninsula microrills are clearly restricted to fine-grained beds. It appears that microrills will develop only on aphanitic rocks in the more humid environments.

We suggest that in all cases gradient controls the plan pattern, but in arid climates the different forms are consequences of differing amounts of effective liquid while in wetter climates the time since exposure to etching processes is most important. In arid climates type 3 microrills develop where the liquid is limited and its movement is governed largely by capillarity. Type 1 is intermediate and reflects more available water. Type 4 is seen most often in the wettest environments. Type 2 is a modified type 1 or 3. In wet climates time is the limiting factor rather than liquid supply. Microrills develop through several initial forms to a final stable form. This is indicated in inland Vancouver Island where etching has been proceeding for varying amounts of time on different surfaces depending on the time since deforestation. On newly exposed surfaces random inhomogeneities of the rock and inter-grain boundaries are etched out to give a type 3 rilling/etching pattern. Type 1 takes over when the lines of drainage have become better established. As solution proceeds further the surface becomes smoother and assumes a type 4 pattern.

In the initial forms flow is controlled largely by capillarity; as etching proceeds, gravity becomes a more important influence but only where the slope inclination is high. The solvent normally will be rainfall, sea spray, soil solution or dew; organic acids may operate locally. The most important requirement is that the rock be fine grained. The grains must be considerably smaller than the rilling, otherwise rock texture will control the form of the surface at that scale. The second requirement is that rain or spray be the only significant erosive agent; biological erosion or frost action, for example, will destroy microrills.

3 RILLENKARREN

Rillenkarren are suites of straight solutional channels displaying remarkable regularity of form and dimension. They develop only on bare rock slopes in some limestones, gypsum and salt. They are the apparent antithesis of normal or Hortonian rills because (a) they head at the crest and diminish in depth downslope where they are replaced by a planar solution surface or "Ausgleichsfläche" (BOGLI 1960) and (b) in any particular occurrence they are of one or two characteristic widths and tightly packed together, rather than being of characteristic widths but uniformly separated by interfluve surfaces (phot.2). They must be the product of direct rainfall (as in Hortonian rills) because in most examples there is no other feasible source of water. They are not seen on very gentle slopes, and on the steepest slopes they break down to cockling patterns (scallops—see below).

Detailed field studies have been reported from Australia (LUNDBERG 1976, 1977, DUNKERLEY 1979, 1983), Canada (GLEW & FORD 1980) and

Photo 2: *(A) Rillenkarren on a limestone block in a landslide pile of Holocene age, Rocky Mountains, Canada. The pencil lies on the lichen-covered Ausgleichsfläche. (B) Rillenkarren on bedrock limestone pinnacles, superimposed on the sides of larger channels, Chillagoe, Queensland, Australia.*

Location	Mean Length (in cm)	Mean Width (in cm)	Rock type	Source
Central Europe	<50	2–3	Various limestones	BOGLI (1960)
Europe	11.8		Various ls.	HEINEMANN et al. (1977)
Yugoslavia	16.0	1.70	Medium cryst. ls.	LUNDBERG (unpub.)
Rocky Mts. Canada		1.25–1.45	Medium to fine, crystalline ls.	GLEW & FORD (1980)
Kalkalpen, Austria	13.0			JENNINGS (1982)
New Zealand	20.0		Coarse-grained ls.	JENNINGS (1982)
Transvaal, Africa		1.76	Dolomitic ls.	MARKER (1985)
From Australia:				
Cooleman Plain	11.6	1.60	Coarse-grained ls.	DUNKERLEY (1979)
Wee Jasper	34.8	1.90	Fine-grained ls.	DUNKERLEY (1979)
Yarrongobilly	41.0		Fine-grained ls.	JENNINGS (1982)
	20.0			JENNINGS (1982)
Chillagoe	22.5	1.74	Recrystallised, coarse ls.	DUNKERLEY (1983)
		1.52	Recrystallised, coarse ls.	LUNDBERG (1977)
	33.5	2.08	Sparry fossiliferous ls.	DUNKERLEY (1983)
		1.95	Sparry fossiliferous ls.	LUNDBERG (1977)
Spain		2.02	Natural salt outcrop	LUNDBERG (unpub.)
Laboratory simulation	14.0	0.55	Plaster of paris	GLEW & FORD (1980)

Table 2: *Summary of Rillenkarren measurements.*

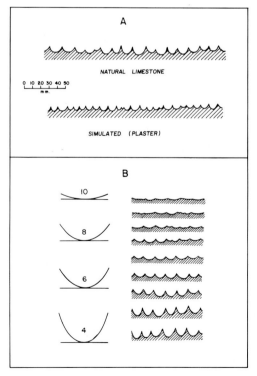

Figure 2: *(A) Rillenkarren cross sections from an experimental plaster block at 60° and from a natural limestone block at 70°, Rocky Mountains, Canada; (B) Evolution of plaster surfaces at 45°, 120–150 hours after beginning dissolution experiment. Parabolas of varying value of n are shown for comparison.*
(from GLEW & FORD 1980)

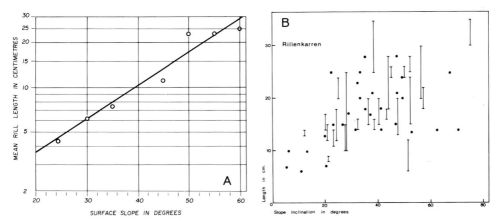

Figure 3: *(A) The log-linear relationship between slope and length from all the plaster experiments of GLEW & FORD (1980); (B) Relationship between slope and length for 80 sites on landslide blocks of crystalline limestones in the Rocky Mountains, Canada, and the Istrian Peninsula, Yugoslavia.*

the German Federal Republic (HEINEMANN et al. 1977). GLEW (1976) and GLEW & FORD (1980) also conducted a laboratory simulation using plaster of paris and a Bryan type rainfall simulator (fig.2). Tab.2 is a summary of Rillenkarren dimensions; the most important observation is that there is a significant range of mean lengths but a great similarity of mean widths on limestones. Rills on natural gypsum and salt probably also have their characteristic uniform widths but there are insufficient data for comparison. GLEW & FORD found the narrowest rills to be on plaster of paris (gypsum), rills were of intermediate width on limestones and widest on salt. Rillenkarren are rare on dolomites; widths appear similar to those on limestones but lengths are shorter.

DUNKERLEY (1983) suggests that chemical considerations would demand about 2600 years for the development of Rillenkarren in limestones. In their gypsum simulations GLEW & FORD (1980) obtained stable equilibrium rills in 250–350 hours.

Relationships between slope angles and lengths have been reported in many papers but conclusions are quite varied. SWEETING (1972) writes that the best developed rills are on 60°–80° slopes but that their length depends on the length of the exposed slope. Most workers have suggested that length increases with angle of inclination. However, HEINEMANN et al. (1979) reported that the longest forms appeared at around 60° with another secondary group forming on domed bosses at around 25°. GLEW & FORD's experiments yielded a strong log linear relationship between length and slope angle (fig.3A). Measures from limestones in the Rocky Mountains and Slovenia show a simple direct relationship between the length and the angle of inclination, but with much dispersion about the trend (fig.3B).

Many authors suggest that Rillenkarren are significantly longer, wider and deeper in tropical or warm areas than in temperate to cold regions. There is some empirical support for this but it is not conclusive. If it exists, we presume the relationship to be an effect of reduced viscosity.

The model of GLEW & FORD (1980) which is explained below suggests that droplet size should be of importance. SWEETING (1972) attributed widespread development of Rillenkarren along the Adriatic coast to intense winter rainfalls experienced there and suggested that their scarcity in the British Isles may relate to the characteristic low intensity, "drizzling" rain. She then presents what might be a more plausible reason; that abundant lichen and other plant growth on the bare rock surfaces minimise rill formation. However, HEINEMANN et al. (1979) found Rillenkarren in association with endolithic algae. At an excellent display in Surprise Valley, Rocky Mountains, Canada, epilithic lichen on the rock surfaces have little effect on the sharpness of the Rillenkarren.

Texture is an extremely important variable. Undoubtedly the best development occurs where texture is uniform, and where the grainsize is medium to fine. Homogeneous gypsum and salt will always display Rillenkarren if the outcrop exposure is stable. Although they sometimes do occur on coarse-grained rock they tend to be rare. They are also rare on dolomites, which are generally coarse-grained and rough. Heterogeneity prohibits rilling where there is a wide range of grain sizes, where insolubles such as siliceous fossils exist or where there is high vug porosity. Rills are also rare where rocks disintegrate into crystal grus (SWEETING 1972).

A direct relationship between rill width and grain size was suggested by GLEW & FORD (1980) on the basis of their simulation results (narrow rills on fine-grained, homogeneous gypsum and wider ones on coarse-grained, homogeneous salt). However the comparison is not strictly valid because of the contrasted chemical compositions. A better test would be to simulate rills on material of the same composition with only crystal size as a variable. DUNKERLEY (1979, 1983) argued for an inverse relationship between rill length and grain size; shorter rills developed in coarse-grained Cooleman Plain limestones and longer forms in nearby but finer-grained Wee Jasper rock, and again for similarly contrasting lithologies at Chillagoe. However LUNDBERG (1976, 1977) suggested that the smaller forms that occur on coarsely crystalline limestones at Chillagoe are simply younger because they are continually being destroyed by granular disintegration and exfoliation.

Effects of other lithological properties are unclear. MARKER (1985) observed that rills in the Transvaal are restricted to certain lithologies (thus proving that rainfall intensity cannot be the primary control) but she failed in an attempt to find any significant relationship with porosity. There were indications that a uniform texture might be of greater importance.

There have been two fundamentally different proposals for the genesis of the form, both seeming to explain its restriction to crests. The first is based upon chemical controls of the reaction kinetics and the second upon physical controls of them.

BOGLI (1960, 1980) proposed that Rillenkarren result from simple molecular dissolution of $CaCO_3$ in reaction with already hydrated CO_2.

$$\underset{calcite}{CaCO_3} + \underset{in\ raindrop}{H_2CO_3} = Ca^{2+} + 2HCO_3^-$$

This dissociation is very rapid and is

completed within a few seconds. The next stage, producing Rinnenkarren, is the hydration of further CO_2 from the rainwater.

$$CO_2 \text{ aqueous} + H_2O \text{ on surface} = H_2CO_3$$

This process requires significantly longer time, so that there will be a band of no erosion between the Rillenkarren and the Rinnenkarren.

The essence of this model is that the instantaneously available reactants, H_2CO_3 or H^+ in the raindrop, are maximum at the crest or any other first point of impact (which must be true) and become significantly exhausted down the length of the rill (which may not be true). The Ausgleichsfläche is then, in Hortonian terms, a belt of least dissolution where solvent capacity is being renewed by further uptake of CO_2 to restore equilibrium in the system, H_2O-CO_2-$CaCO_3$.

An important question thus is the rate at which the Ausgleichsfläche is lowered in comparison with the rills. BOGLI's chemical model requires little erosion beyond the base of the rills (see fig.4A). However, his own diagram suggests that this is the site of maximum erosion (BOGLI 1980 and fig.4B). GLEW & FORD (1980) measured rates of lowering on their simulation surfaces by micrometer, using a Kelvin clamp arrangement, and demonstrated that both the rills and the planar slope are being removed at about the same rate (fig.4C—it is of interest that this profile is the most commonly observed in the field).

The fact that natural Rillenkarren with Ausgleichsflächen also develop on gypsum and salt confutes this chemical proposal for their development on limestone. Gypsum and salt dissolve by simple dissociation in the presence of water and there are no further reaction steps such as CO_2 uptake to retard or accelerate the process. In their simulation GLEW & FORD found that the solution discharged from their models was less than 1% saturated. Both Rillenkarren and Ausgleichsfläche can be found where the solution remains very aggressive with respect to the soluble mineral.

The simulations demonstrated that Rillenkarren propagate from the crest downslope until a stable length is reached and that both Rillenkarren and the Ausgleichsfläche are then removed by simple parallel retreat. Rill formation begins as many short, shallow rills at the crest; they deepen and lengthen, coalescing laterally and thus consuming smaller "hanging" rills. Rills at the lateral edges of surfaces are longer because thin flow is maintained here. The rill cross section approaches a parabolic form at equilibrium (fig.2). The rate of surface lowering depends on the intensity of the rainfall and on the gradient (the maximum rate was at around 45° in the experiments). GLEW & FORD (1980) proposed a physical "rim-effect" model. At the crest or rim raindrops penetrate right through the saturated, viscous sublayer of film flow, permitting turbulent reaction at the rock surface. Downslope from the crest the depth of flow increases to some critical value where the drops can no longer impact the surface directly; i.e. the rim effect no longer applies. Rillenkarren are the characteristic stable forms within the zone of rim effect. The parabolic cross section is interpreted by GLEW & FORD as the most efficient shape to minimise droplet impact on channel sides and thus concentrate energy into the central portion of the channel. DUNKERLEY (1983) describes the form as hyperbolic. It does not really matter; any concave form will focus the energy.

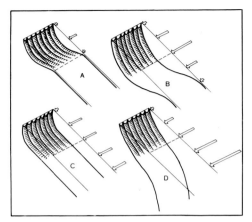

Figure 4: *The development of the rilled section and the Ausgleichsfläche under various solution regimes; the lengths of the arrows represent the differing rates of solution, the original surfaces are shown in faint lines, the resultant surfaces in bold.*

4 SOLUTION RUNNELS (RINNENKARREN AND RUNDKARREN)

4.1 RINNENKARREN

Rinnenkarren are the equivalent on soluble bare rock of Horton's first order rills. The "Ausgleichsfläche" becomes "the belt of no (channelled) erosion". This is the zone, where discharge builds up until the sheet flow mode breaks down into linear streams. A series of shallow, sub-parallel rills develops perpendicular to the slope, as in Horton's belt of rill erosion. The diagnostic characteristics of Rinnenkarren are that they head below the Ausgleichsfläche, are separated by distinct interfluves, are much bigger than Rillenkarren, increase in width and depth downslope, and have sharp channel rims and rounded bases. JENNINGS (1985) observes that the interfluves separating them may be substantial (several times the mean Rinnenkarren width) and may carry Rillenkarren. The lower boundary of the Ausgleichsfläche is often imprecise and hard to identify: because no threshold flow velocity is required for solution to occur, at the base of the Ausgleichsfläche there exists an area in which channel formation is gradually becoming established rather than a clear channel head line as in many true Hortonian rills on soils. Typical measures we have taken in northern Vancouver Island show Rinnenkarren increasing in width from about 4 cm to 8 cm, and in depth from about 2 cm to 7 cm, with lengths of around 5 m. SWEETING (1972) quotes depths up to 50 cm and lengths up to 20 m for Rinnenkarren elsewhere.

Rinnenkarren are common in the Alps, the Dinaric Karst and other mountainous regions where bare limestone slopes (normally bed surfaces) occur. There are spectacular forms in the tropics; SWEETING quotes examples from Sawarak and the Fitzroy area of Australia where the interfluves approach knife-edge forms, as in Rillenkarren (however, see discussion in paragraph below).

BOGLI (1960) classified Rinnenkarren as a primary form attributable to uniform rainfall coverage, and caused by dissolution that consumes the CO_2 that dissolves and complexes in the rainwater before longer term re-equilibration with the atmosphere becomes the rate control. Heavy runoff rarely reaches chemical equilibrium on steep rock slopes so that Rinnenkarren potentially can be very long (many tens of metres). The form may be slightly sinuous on gentle slopes but it becomes straighter with increasing slope. There are few junctions and the junction angles are very acute. "Regenrinnenkarren" are a straight subtype found on the steepest slopes.

"Meanderkarren" and "Wandkarren" (see below) may be confused with true Rinnenkarren because they have similar scales and channel form. According to BOGLI (1960) their cross sections shallow downslope, but it is clear that in many cases "Meanderkarren" are simply Hortonian rills which happen to meander more than is common, whereas in other cases they are true decantation forms as discussed below. SWEETING (1972) includes them under "Rinnenkarren" as forms which may be either completely free (bare rock) or half-free (covered to some extent by vegetation or soil).

BOGLI (1980) retained his basic distinction of 1960 between forms which deepen and those which shallow downslope. However JENNINGS (1985) included all the forms of this size class (Rinnenkarren, Meanderkarren and Wandkarren) under "forms developed on bare karst with concentrations of runoff" without regard to their downslope development. BOGLI's division is possibly the more correct since it emphasises the different chemical and hydrological regimes. "Rinnenkarren" should be restricted to rills resulting from the breakdown of surface sheetflows.

In our experience, many exposed Rundkarren (see below) resemble Rinnenkarren, and thus may be misidentified, especially if the sides have been sharpened by subaerial weathering. If the history of former soil/vegetation cover is considered then it often becomes clear that the forms must have originated under cover, i.e. are in fact Rundkarren. In the literature the distinction is not always clear. True Rinnenkarren form subaerially throughout their development, by definition. Many of the larger runnels of the tropics were established under cover and later exposed to subaerial sharpening; examples from the Chillagoe tower karst of Australia and from the Stone Forest of China show this clearly.

This review of Rinnenkarren has highlighted the need for careful identification and the paucity of field-based morphometric analyses of these features. There is scope for much further research.

4.2 RUNDKARREN

Rundkarren are rounded solution runnels which are generally observed where a soil cover has been removed (phot.4). They are similar in size to Rinnenkarren, being about 12–50 cm wide, and may deepen downslope. In the literature morphometric analyses of Rundkarren are rare. Fig.5 presents some measures from Vancouver Island. The length of runnel which can be measured depends on the length of the slope and the extent of soil removal from it; most were exposed to lengths of around 3 m. Depths typically increased downslope from 1 cm to around 20 cm, widths from 2 cm to 14 cm.

Rundkarren walls are smooth and much more rounded at the shoulder than those of Rinnenkarren. They often display regular dendritic drainage patterns, but on steeper slopes will tend to be sub-parallel (SWEETING 1972). The channel floors commonly have sections

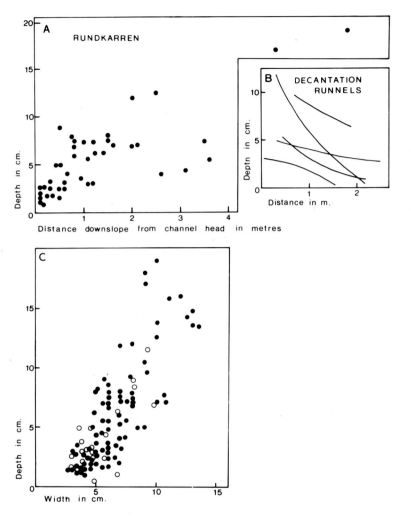

Figure 5: *Morphometric analysis of Rundkarren and decantation runnels from northern Vancouver Island, Canada*
(A) Depth changes downslope from the channel head in Rundkarren; (B) Depth changes downslope in Decantation Runnels; (C) Width: Depth relationships for both types of runnels show a similar behaviour, suggesting why they are often confused (Rundkarren are shown as solid circles, Decantation Runnels as open circles).

of reversed slope or distinct depressions within them. The purest Rundkarren (i.e. not polygenetic) have formed beneath a complete cover, perhaps of soil, of acid till, of moss or of litter. The strong shoulder rounding is attributed to lateral soil-base flow causing solution roughly normal to the channel orientation. This is unlikely to occur on bare slopes. Similarly, the presence of reversed slopes and basins within the channel argues for development under a cover where simple gravitational draining is checked. However, when it is very wet there is no doubt that regular channel flow does occur in the Rundkarren bases washing out some of the soil.

On flat surfaces of limestone and dolomite (clint blocks) Rundkarren develope accordant to hydraulic gradients in the clastic cover that are oriented to earlier solutional openings along major joints (grikes). Where a fissure or junction of such fissures occurs water can more easily drain so that a drawdown cone is created in the cover. Water flows faster down the cone and induces piping or percolines on the rock surface, so initiating centripetally arranged Rundkarren. Where a joint is uniformly open along its length the spacing of Rundkarren draining into it is related to the permeability of the cover, to the proximity of other open joints (because of competition for drainage) and to the slope. Centripetal patterns are most likely to occur where joint frequency is low and where bedrock and soil surface slopes are very gentle. In Yorkshire, England, JONES (1965) showed that increasing slope angle causes a change from dendritic to parallel patterns and that the extent of Rundkarren incision is related to the amount of time that the rock was covered; the most recently emergent surfaces (i.e. which were covered for the longest time) have the deepest forms and the steepest long profiles.

Newly exposed Rundkarren display the characteristic smooth rounded surfaces of subsoil solution. When they have been exposed for some time, the edges become sharpened and their subsoil origin is harder to identify. In reality many sites go through a history of vegetational changes, cover changes, etc. so that a clear distinction between bare and covered karst is impossible. In Vancouver Island virgin forest areas display patches of rock which are currently bare, perhaps where a tree has fallen or a shaft is open in the rock beneath the tree roots. In this case the Rundkarren may go through alternating buried and bare stages. This concept of overlapping forms and mixed origins is important and probably explains the majority of the runnel forms currently exposed in the Karst areas of the world. Removal of cover through human intervention has clearly occurred in many parts of the world and the time since exposure has not been sufficient to permit formation of the deep runnels commonly observed; these must have been moulded by subsoil solution and have merely been modified by subaerial solution, exfoliation or frost action. This is clearly the case for most of the well known European Rundkarren pavements where the land has been used continuously and intensively since clearance.

Rundkarren are more common than true Rinnenkarren but the two are often closely related; for example, where the crest of a slope is bare but lower parts are soil mantled, the pattern is of Rinnenkarren feeding to Rundkarren, the two developing concurrently, their width, spacing and other proportions similar. BOGLI (1960) hinted at the possible interactions by suggesting that in humid climates any bare limestones are eventually colonised by vegetation so that original subaerial karren become covered. The vegetation and soil cover then cause rounding of the shoulders. Conversely, SWEETING (1972) argued that the celebrated runnels of Hutton Roof in Britain (phot.3A), which are narrow, have rounded edges and deepen downslope, are basically rundkarren which have been modified since emergence from under the soil so that they now resemble closely spaced, overdeepened Rinnen. Modern clearance of virgin forest in some areas of Vancouver Island has led to soil erosion and exposure of Rundkarren very similar to those of Hutton Roof (phot.3B).

Past classifications have placed Rinnenkarren and Rundkarren in distinct, separate categories. However, the field evidence indicates that there is a wealth of interaction so that clear generic distinction may not be possible at a given site. It certainly appears unnecessary.

5 DECANTATION OR OVERSPILL FORMS

Decantation or overspill forms includes "partly covered" or "halbfrei" karren of previous classifications.

The dissolutional rill forms comprising this group are much more varied in their morphology, scale and distribution than are Rillenkarren or Rinnenkarren. In contrast to previous workers we place them into a single category because they share one fundamentally important feature—the solvent is supplied as an overspill from an upslope store rather than directly from rainfall. It flows down a bare rock slope into which the rill is then indented below the overspill point. The store may be soil, moss, humus etc. retained in a depression and overspilling to form Wandkarren on steep slopes or Meanderkarren on gentle slopes; it may be a snow bank; it may be a very ephemeral store such as tiny rainpits at the crest of Rillenkarren slopes which overspill to enlarge particular rills below them. In the case of decantation flutings (typically on the walls of grikes or of dome pits in caves) the store is often an exposed bed surface or a bedding plane. There may be some enhancement of solvent aggressivity during the storage. The diagnostic characteristic of decantation forms is that, because they may not collect extra water downslope, the deepest part of the rill is at or close to the overspill point; it becomes shallower downslope and will eventually extinguish if the available slope is long enough. In reality, decantation and direct rainwater forms tend to be intermixed; e.g. a Rinnenkarren may contain a store of moss/soil. Immediately downslope of the store the overdeepened rill is formed both by rainwater and by decantation.

Decantation forms are of two basic types. Where overspill is from a point, then a single, sharp-edged rill will develop, its scale related simply to the size of the store and the aggressivity of the solvent and its form related to the hydraulics of the flow. These are the "Wandkarren". Where overspill is in a diffused or sheet form, then a suite of closely packed rills with sharp interfluves will form. These are similar to Rillenkarren but usually wider and of shallower proportion. These are the decantation "Flutings".

5.1 WANDKARREN OR WALL KARREN

The smallest member of this family is the extended, deepened rill caused by overspill from crest rainpits; the largest are transitional to regular stream channels which enlarge downslope because of their large catchments and the introduction of abrasion by bedload, etc. Large examples on steep dipslopes in the Rocky Mountains of Canada are 100 m or more in length and 50–80 cm wide and deep. Typically they are medium to large rills which superficially resemble Rinnenkar-

Photo 3: *(A) Runnels from Hutton Roof, England; (B) Runnels from Vancouver Island, Canada.*

Photo 4: *Decantation forms: Wandkarren from Norway.*

ren (phot.4) but may be distinguished in that they tend to become shallower and to extinguish downslope.

Most Wandkarren are straight and oriented downslope but meandering, sinuous to highly sinuous forms are sometimes observed. Meandering Wandkarren occur on two scales. The larger, more common forms occur on slopes up to about 20° inclination. These are on the Rinnenkarren scale and meander because the gradient is comparatively low. The second type (phot.5) develop on slopes up to and including the vertical. They are much smaller, up to 1 cm wide, and can be considered transitional from the micro to the small scale of karst phenomena. Here we suggest that the fluid is released so slowly that flow is retarded by surface tension effects and deflected by rock inhomogeneities, grain boundaries, etc., causing the tight meandering,

5.2 DECANTATION FLUTINGS

Decantation flutings (a class not previously defined) develop where there is a steady supply of water to near-vertical to overhanging slopes and the resulting film flow can occur with no interruptions from raindrop impact, wind, biological activity, etc. Once again the forms display the best, most regular development where rocks are comparatively homoge-

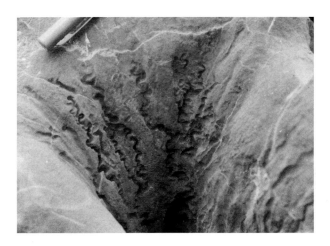

Photo 5: *Micromeanderkarren from Vancouver Island, Canada: these have originated from very slow decantation underneath the former natural forest cover.*

neous and fine grained. These flutes are common in caves where flow from horizontal bedding planes discharges down the walls of vertical, joint-guided shafts. Dimensions are typically 5–25 cm in width in such cases, and lengths can exceed 25 m. In the ideal case there is uniform decantation around the rim of a shaft so that its walls are indented by packed (adjoining) flutes of uniform width. Normally there is some variety of width (greater than occurs in mature Rillenkarren patterns, for example) and some parts of the shaft circumference are without flutes. This is due to spatial irregularities in the decantation. Excellent flutings are observed on the walls of deep grikes in Vancouver Island, Canada. These are around 10 cm wide and regularly packed.

DUNKERLEY (1983) mentions flutes developed under overhanging ledges in calcareous aeolianite in Western Australia where film flow from joints occurs. Decanting flutes occur at the base of sea cliffs inCo. Clare, Ireland, where a film of water remains after each passing wave. Flutes are reported in basalt in Hawaii that also appear to be of the decanting type; PALMER (1927) observed that they formed where water emerged from horizontal joints.

The mechanism for formation requires that discharge be adequate to maintain flow attachment, but not so great that cavitation can occur. Where the slope exceeds approximately 70° Rillenkarren surfaces breakdown to the roughly linear scallop patterns termed "cockling" by SWEETING (1972). Therefore, it must be presumed that the fluting film is thinner than that causing Rillenkarren. On plane surfaces of wide extent perturbations occur transverse to the primary flow that cause separation into parallel opposing vortices or "Parting Lineations" (ALLEN 1970). Parting lineations are a universal feature of the laminar sublayer and lower buffer layer in the turbulent boundary layer. If this mechanism causes the formation of decanting flutes then their width will be inversely proportional to the velocity of flow. Velocity and depth of flow are constant in the ideal case because the source bedding plane or joint can only transmit water at a constant rate.

6 POLYGENETIC FORMS

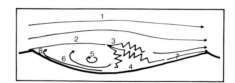

Figure 6: *Hydrodynamics of fluted scallops (after BLUMBERG & CURL 1974): Hydrodynamic regions for flow over one flute period.*
1. Main turbulent flow.
2. Laminar free shear layer.
3. Transition to turbulence with flow separation.
4. Reattachment.
5. Irregular recirculating flow.
6. Lee-slope boundary layer.
7. Streamward-slope boundary layer.
8. Lee-slope boundary layer separation with small eddies.

SAURO (1973, 1975) emphasises that it may be extremely difficult to distinguish the genetically different forms of runnel in the field, especially on gentle slopes. Some may have begun as simple overland flow forms but have been invaded by clumps of vegetation that created deepened and/or widened pan-like sections, as in Hohlkarren discussed below. Others have begun as solution pans (Kamenitza) that overspilled to form Meanderkarren. Many of the apparently simple Rinnenkarren have been triggered

Photo 6: *Development of Pinnacles in the celebrated Stone Forst of Yunnan, China: (A) Rounded surfaces emerging from soil cover and being sharpened by subaerial solution; (B) The end product.*

by localised biological activity. These we have classified as Polygenetic forms.

Hohlkarren are intrusions into gravitomorphic forms from filling by humus. BOGLI (1960) classifies them as secondary, semi-free forms. Biogenic CO_2 acts to steepen and undercut the walls, creating sack-like cross sections. The aggressivity will be used up so that the grooves tend to flatten out 2–3 m downslope of the humus store. There are many gradations between Rinnenkarren, Meanderkarren and Hohlkarren.

Pinnacles represent the residual forms where extreme development of karren has removed much of the rock. They appear to be attributable to initial subsoil dissection by Rundkarren of clint blocks in very thick bedded to massive strata. The pyramidal forms thus created become exposed and cannot easily retain litter or soil; they are then sharpened by free rill forms. BOGLI (1980) suggests that pinnacles form where the side walls of grikes and Rinnenkarren cut across one another to form sharp edges and peaks which can reach several metres in height.These need a long time to develop and so they are rare in recently glaciated areas. They are common in the humid tropics. The greatest displays are at the "Stone Forest" of Yunnan, China (CHEN ZHI PING et al. 1983) where the progression from subsoil rounded pinnacles to subaerial sharpened pinnacles is clear (phot.6).

7 RILLS TRANSVERSE TO THE DIRECTION OF FLUID FLOW

This final category of linear dissolution forms is quite distinct from the others and they are not gravitomorphic. Fluted scallops occur underground and horizontal ripples above ground. Both are rare.

Fluted scallops develop on cave passage walls and ceilings where solution occurs under conditions of complete waterfill i.e. phreatic or epi-phreatic (floodwater) conditions. They superficially resemble transverse ripples in sand. They are oriented normal to the direction of flow and have an assymetric cross section, being steeper on the upstream side. They are packed, one abutting on to the next. They are recognised as ideal end members of the class of erosional scallops.

The theory for scallops was developed by CURL (1966) and has been confirmed by several sets of flume experiments (ALLEN 1971, GOODCHILD & FORD 1971, BLUMBERG & CURL 1974). On a dissolving or ablating surface in a fluid flow with a Reynold's Number $\geq 22,000$, boundary-layer detachments occur in the lee of prominences. Dissolution rate is enhanced at the points of reattachment. This normally produces scallops, scooped depressions like the bowls of spoons, that are oriented downstream. The length of the scallops is proportional to fluid viscosity and inversely proportional to its velocity. Where the turbulent flow possesses great lateral stability the scallops widen into flutings transverse to the flow. Length transverse to the flow may be many metres. Length in the direction of flow is 2–10 cm. In theory they should migrate downstream as sand ripples do, but this is not confirmed. Such fluted scallops are very rare features; they develop only where the dissolving rock is sufficiently homogeneous to permit lateral extension of the trough and the flow regime is very stable. Fig.6 shows the essential hydrodynamic features.

Ordinary solutional scallops are also common on the surface. As noted above, Rillenkarren break down to scallop patterns on limestone surfaces steeper than approximately 70°. This "cockling" develops under high velocity shallow flows. Only where the flow is stable will the scallops tend towards parallel, horizontal (transverse) ripples. JENNINGS (1985) shows "solution ripples" on marbles of New Zealand. WALL & WILFORD (1966) describe very clear horizontal ripples from Sarawak. JENNINGS (1985) suggests that there are two kinds of solution ripple: the first, exemplified by those in Sarawak, are around 2 cm across and 1 cm deep, occur on underhangs, may continue horizontally for tens of centimetres, and have sharp ribs separating the flutes. The flow has probably acquired additional aggressivity from soil or humus. The second type, e.g. in New Zealand, are shallower, less symmetrical, with blunter edges. These occur on the more gentle slopes and become step-like.

8 CONCLUSIONS

A surprisingly wide range of linear to slightly sinuous, purely dissolutional, phenomena develop on the more readily soluble rocks, especially on limestones. A variety of different genetic mechanisms or controls must be called upon to explain them. In only two cases (Rillenkarren and scallops) is there direct experimental evidence from simulations such as are the standard practice in rill work on soils. Rates of development, even on plaster of paris, are so low that sophisticated hardware experiments with e.g. covered forms, would probably take far too long. There is scope for more inferential work and theoretical genetic modelling.

The smallest scale forms (microrills) appear to be controlled by capillarity. At the small scale, on bare rock, both Rillenkarren and flutings appear to be boundary layer phenomena, although of radically different kinds. At high gradients, rilling breaks down to scalloping as the saturated boundary detaches. Rinnen- and Rundkarren are more normal rilling phenomena in terms of their spacing (separation and distribution). Their pattern changes from dendritic to parallel as a function of slope. Their fundamental morphometric relations and catchment dynamics have not been studied adequately. Decantation forms have spacing and dimensions determined by characteristics of the storage reservoir. Wall flutings in dome pits may be distributed by an edgewave function. Wandkarren on open bedrock slopes include the longest and deepest dissolutional rills and become transitional to regular stream channels that happen to be incised into especially soluble rocks.

REFERENCES

[ALLEN 1970] ALLEN, J.R.L.: Physical Processes of Sedimentation. Unwin Univ. books, London.

[ALLEN 1971] ALLEN, J.R.L.: Transverse erosional marks of mud and rock: Their physical basis and geological significance. In: Sedimentary Geology, Intern. J. Applied Regional Sed., Elsevier, Amsterdam.

[BLUMBERG & CURL 1974] BLUMBERG, P.N. & CURL, R.L.: Experimental and theoretical studies of dissolution roughness. J. Fluid Mechanics, 65, 735–751.

[BOGLI 1960] BOGLI, A.: Kalklösung und Karrenbildung. Z. Geomorph., Suppl. 2, 4–21.

[BOGLI 1980] BOGLI, A.: Karst hydrology and physical speleology. Berlin, Springer-Verlag, 284 pp.

[CHEN et al. 1983] CHEN ZHI PING, SONG LIN HUA & SWEETING, M.M.: The pinnacle Karst of the Stone Forest, Lunan, Yunnan,

China: an example of a sub-jacent karst. In: PATERSON,K. & SWEETING, M.M. (Eds.): New directions in Karst. Geobooks, Regency House, **33**, 597–607.

[CURL 1966] CURL, R.L.: Scallops and Flutes. Trans. Cave Res. Gp. Gt. Brit., **7**, 121–160.

[DUNKERLEY 1979] DUNKERLEY, D.L.: The morphology and development of Rillenkarren. Z. Geomorph., **23**, **3**, 332–348.

[DUNKERLEY 1983] DUNKERLEY, D.L.: Lithology and micro-topography in the Chillagoe Karst, Queensland, Australia. Z. Geomorph., **27**, **2**, 191–204.

[FORD 1980] FORD, D.C.: Thresholds and limits in karst geomorphology. In: FREDERKING, R.L. (Ed.): Thresholds in Geomorphology. Dowden and Hutchinson Co., New York. 345–362.

[GLEW 1976] GLEW, J.R.: The simulation of Rillenkarren. Unpublished McMaster University M.Sc. thesis, 116 pp.

[GLEW & FORD 1980] GLEW, J.R. & FORD, D.C.: A simulation study of the development of Rillenkarren. Earth Sur. Proc., **5**, 25–36.

[GOODCHILD & FORD 1971] GOODCHILD, M.F. & FORD, D.C.: An analysis of scallop patterns. J. of Geology, **79**, 52–62.

[HEINEMANN et al. 1977] HEINEMANN, U., KLAADEN, K. & PFEFFER, K.H.: Neue Aspekte zum Phänomen der Rillenkarren. Abh. Karst und Höhlenkunde, Reihe A Spelaologie, **15**, 56–80.

[JENNINGS 1981] JENNINGS, J.N.: Morphoclimatic control—a tale of piss and wind or a case of the baby out with the bathwater? Proc. 8th Int. Spel. Cong., **1**, 367–368.

[JENNINGS 1982] JENNINGS, J.N.: Karst of northeastern Queensland reconsidered. Tower Karst, Chillagoe Caving Club, Occ. Pap. **4**, 13–52.

[JENNINGS 1985] JENNINGS, J.N.: Karst Geomorphology. Blackwell, Oxford, 293 pp.

[JONES 1965] JONES, R.I.: Aspects of the biological weathering of limestone pavements. Proc. of the Geologists Ass., **76**, 421–434.

[LAUDERMILK & WOODFORD 1932] LAUDERMILK, J.D. & WOODFORD, A.O.: Concerning Rillensteine. Am. J. Sci., **223**, Series 5, 23, 135–154.

[LONGMAN & BROWNLEE 1980] LONGMAN, M.W. & BROWNLEE, D.N.: Characteristics of karst topography, Palawan, Phillipines. Z. Geomorph., **24**, 299–317.

[LUNDBERG 1976] LUNDBERG, J.: The geomorphology of Chillagoe limestones: variations with lithology. Unpublished M.Sc. thesis, Aust. Nat. Uni., Canberra, 175 pp.

[LUNDBERG 1977] LUNDBERG, J.: An analysis of the form of Rillenkarren from the tower Karst of Chillagoe, Australia. Proc. 7th. Int. Spel. Cong., Sheffield, England, 294–296.

[MARKER 1985] MARKER, M.E.: Factors controlling micro-solutional karren on carbonate rocks of the Griqualand West sequence. Cave Science, **12**, **2**, 61–65.

[PALMER 1927] PALMER, H.S.: Lapies in Hawaiian basalts. Geogr. Rev., **17**, 627–631.

[PERNA & SAURO 1978] PERNA, G. & SAURO, U.: Atlante delle microforme di dissoluzione carsica superficiale del Trentino e del Veneto. Memorie del Museo Tridentino di Scienze Naturali, XXII, Nuova serie, Fasc. unico, 1–176.

[SAURO 1973] SAURO, U.: Forme di corrosione su rocce mononate nella Val Lagarina meridionale. L'Universo, **53**, **2**, 309–344.

[SAURO 1975] SAURO, U.: The geomorphological mapping of "Karrenfelder" using very large scales: an example. In: Karst Processes and relevant landforms. Ed. Gams, I., 189–199.

[SWEETING 1972] SWEETING, M.M.: Karst Landforms. 362 pp. London, (Macmillan)

[TRUDGILL 1973] TRUDGILL, S.T.: Limestone erosion under soil. Proc. 8th. Int. Cong. Spel., **2**, Ba 044, 409–422.

[TRUDGILL 1985] TRUDGILL, S.T.: Limestone Geomorphology. Longman, London and New York.

[WALL & WILFORD 1966] WALL, J.R.D. & WILFORD, G.E.: A comparison of small-scale solution features on microgranodiorite and limestones in West Sarawak, Malaysia. Z. Geomorph., **10**, 462–468.

Address of authors:
D.C. Ford and J. Lundberg
Dept. of Geography,
McMaster University
Hamilton, Ontario L8S 4K1,
Canada

RILL DEVELOPMENT AND BADLAND REGOLITH PROPERTIES

J. **Gerits**, A.C. **Imeson**, J.M. **Verstraten**, Amsterdam
R.B. **Bryan**, Scarborough

SUMMARY

Rill systems often attain their most conspicuous development on badland slopes in semi-arid regions. This paper considers the characteristics of rill systems developed on Eocene and Miocene marls at two sites in Granada Province, southeastern Spain, and on smectitic mudstones in western Canada. Rill system development is closely related to regolith hydrologic properties and is viewed as being analogous to, and governed by, essentially the same properties as artificial field drains. Close similarities in regolith characteristics and response to rainfall in the two areas are noted. Patterns of rill development are strongly influenced by properties such as consistency, shrink-swell capacity and dispersibility, which reflect clay mineralogy, content and chemistry, and the concentration and chemistry of pore water. These properties control regolith hydraulic and hydrologic characteristics. Where these properties are extreme, as at Dinosaur Park, Canada, and Cortijo, Spain, subsurface crack and micropipe flow is highly developed, and leads to rill development. Where these properties are less pronounced, at Dehasas, Spain, such flow is less significant and rill systems are less dense, and more frequently obliterated by shallow mass wasting.

1 INTRODUCTION

Intense rill incision characterizes most badland slopes, but some are conspicuously undissected. In badland areas rill distribution frequently seems to coincide with certain lithological units. This coincidence may have several causes:

1. rilled units may occupy certain critical locations in the landscape;

2. they may coincide with particularly steep slopes;

3. the surface characteristics of weathering products may critically affect the hydraulics and concentration of surface flow.

In this paper an alternative possibility is examined, that in some badland areas rill occurrence is related primarily to properties which control the internal dynamic response of the regolith to wetting by rainfall. The response processes concerned are: flocculation, dispersion, slaking, shrinking and swelling. These affect the hydraulic properties of the regolith during rainfall events, and determine the pathways which water draining

ISSN 0722-0723
ISBN 3-923381-07-7
©1987 by CATENA VERLAG,
D–3302 Cremlingen-Destedt, W. Germany
3-923381-07-7/87/5011851/US$ 2.00 + 0.25

the regolith can follow. The relationship of these processes to rill incision patterns is examined with reference to two badland areas, one in the valleys of the Rio Fardes and the Rio Guadahortuna in southeastern Spain (IMESON & VERSTRATEN 1986, GERITS 1986, 1987), and the other in the valley of the Red Deer River, Alberta, Canada (CAMPBELL 1970, BRYAN et al. 1978).

The Perth Amboy badlands on which SCHUMM's (1956) concepts of rill system development were based, were developed in artificial fill of comparative homogeneity, and most of the more recent papers on rill erosion have focussed primarily on cultivated land. Natural badland rill systems differ from those developed in cultivated or otherwise artificial conditions, and concepts developed for one system are not necessarily relevant for the other. Some badland rill systems develop seasonally in regolith which is freshly exposed by mass-wasting processes, or disturbed by such processes as freeze-thaw activity (SCHUMM & LUSBY 1963). These are analogous to "cultivation" rills in which tillage replaces natural seasonal disturbance. In each case, development of the rill system is typically associated with high rates of sediment production which increases progressively with extension of the incision network until an (equilibrium) maximal extension is reached. Whether or not there is sufficient time to reach this stage between seasonal or tillage disturbances will depend on local rainfall and regolith conditions. Unlike these ephemeral rills, other badland rill systems may persist over many years, and are not necessarily associated with high erosion rates. These may be thought of as systems in a state of dynamic equilibrium (LEOPOLD & LANGBEIN 1962, ABRAHAMS 1968), or they may be relict features produced in climatic conditions which no longer prevail.

The properties of badland regolith materials considered in this paper are not the primary physical, mineralogical or chemical properties which govern material behaviour. These can be found in HODGES & BRYAN (1982), BRYAN et al. (1984) and GERITS (1986, 1987). For convenience, however, some information on chemistry and mechanical composition is given in tab.1 and 2. The main focus is the secondary properties such as consistency, shrinking, swelling and clay dispersion, which influence regolith hydrology and the release of sediment and solutes.

1.1 CONDITIONS FAVOURING RILL INITIATION

In most studies rill initiation has been explained in terms of flow hydraulics, following HORTON (1945) who considered rill development to be related to the velocity and hence the depth of overland flow. More recently field and laboratory experiments on the hydraulics of rill flow in Belgium have demonstrated that rill initiation reflects critical values of the Froude number (SAVAT 1979, SAVAT & DE PLOEY 1982), or the shear velocity (DE PLOEY 1984, GOVERS 1985) and hence the transport capacity of flow. These physical relationships were shown to be valuable in explaining the occurrence of rills in the loess region of Flanders.

Rill initiation can result solely from the hydraulic action of water on essentially inert particles (MEYER & MONKE 1965), but on most natural slopes these physical processes are affected by physico-chemical interaction

Regolith type & horizon	EC_{25} (mS/cm)		SAR_p $(mmol/l)^{1/2}$		$Na/(NO_3+SO_4)$		Na/Cl		Ca/SO_4		% H_2O^-	
	Deh.*	Cort.**	Deh.	Cort.	Deh.	Cort.	Deh.	Cort.	Deh.	Cort.	Deh.	Cort.
Pinnacle												
crust	0.91	2.03	3.6	10.5	2.9	2.0	3.6	1.6	1.5	0.2	44.0	49.6
subcrust	10.58	14.73	11.6	24.6	0.9	3.4	1.6	1.0	0.5	0.7	55.8	64.6
weath. l.	15.11	45.77	19.6	34.5	0.8	1.7	2.3	1.2	0.3	0.4	55.7	58.5
shards	17.21	35.20	40.0	29.7	0.7	1.4	1.6	1.3	0.03	0.3	88.9	59.6
marl	17.57	27.30	81.9	30.3	1.0	1.6	1.5	1.2	0.02	0.2	97.4	59.6
Tower												
crust	4.67	3.25	15.3	14.8	11.6	3.8	1.1	1.2	2.8	0.4	61.7	50.8
subcrust	22.21	15.78	39.0	21.6	13.9	1.9	1.0	1.2	3.3	0.4	83.0	57.0
weath. l.	14.72	25.91	38.3	26.0	6.5	1.8	1.1	1.4	1.0	0.3	149.7	61.0
shards	34.79	28.38	41.9	29.2	6.7	1.5	1.0	1.3	1.8	0.3	82.4	57.5
marl	33.80	14.70	39.8	26.7	7.4	2.4	1.0	1.2	2.1	0.3	89.6	59.9
Domed												
crust	36.02	12.17	37.7	15.2	9.7	2.9	1.0	1.1	4.2	1.2	56.5	48.7
subcrust	49.02	30.20	34.1	17.5	5.1	2.2	1.0	0.9	4.2	1.4	65.2	55.0
weath. l.	-	28.94	-	22.3	-	2.3	-	1.0	-	0.9	-	57.5
shards	58.81	46.47	32.7	24.2	4.8	6.0	0.9	0.8	32.7	6.3	72.6	57.6
marl	n.d.	39.06	n.d.	20.7	n.d.	5.4	n.d.	0.8	n.d.	8.6	n.d.	49.8
Pediment												
crust	14.58	2.05	13.6	4.8	3.4	3.6	0.9	1.0	2.1	1.7	34.6	30.1
subcrust	42.50	16.85	32.3	16.0	2.4	9.9	1.0	0.9	0.8	3.4	33.9	44.7
weath. l.	38.34	15.17	26.2	22.1	2.0	3.0	1.0	1.0	0.9	0.4	45.3	55.7

* Deh. = Dehesas; ** Cort. = Cortijo

Table 1: *Some characteristics of regolith materials from the Dehesas and Cortijo badland areas (saturated pastes extracts).*

Regolith material	EC_{25} (mS/cm)	SAR_p $(mmol/l)^{1/2}$	$Na/(NO_3+SO_4)$	Na/Cl	Ca/SO_4	% H_2O^-
grey shale	7.85	24.2	2.1	>3000	0.20	160.6
sandstone	2.27	22.0	1.3	313	0.06	95.1
upper grey shale	4.74	32.7	-	-	-	119.1
yellow shale (1)	6.24	33.1	1.7	657	0.08	132.0
pediment	6.57	21.0	n.d.	n.d.	n.d.	38.0
yellow shale	4.15	39.0	1.8	460	0.05	164.0
yellow shale 0–2 cm	6.61	31.0	1.7	342	0.10	142.5
yellow shale 2–4 cm	8.65	33.7	1.6	528	0.10	168.3
yellow shale 4–8 cm	6.21	32.3	1.8	n.d.	0.10	283.9
yellow shale >10 cm	5.63	34.5	1.7	>3000	0.08	160.5

Table 2: *Some characteristics of the regolith materials from the Dinosaur Park Badlands (saturated pastes extracts).*

between the soil/regolith and the eroding water. These can strongly influence detachment, transport and deposition of cohesive materials (GRISSINGER 1966, PARTHENIADES 1972, GERITS 1987). For example, the erosion rates of cohesive materials are dependent upon the chemistry of pore fluid and eroding water (SARGUNAM et al. 1973, ARIATHURAI 1978, ARULANANDAN et al. 1978) so that the critical shear stress required to initiate erosion will vary during a rainfall event in response to changes in both runoff hydraulics and chemistry. Consequently, entrainment of cohesive materials can, within certain limits, vary independent of the fluid shear velocity. Furthermore, the hydraulic characteristics of the flow (shear velocity and shear stress) appear to be influenced by physico-chemical processes in the boundary layer between the fluid and the cohesive material (GERITS 1987). Chemical properties which are indices of the effect of water chemistry on erosion are the sodium adsorption ratio (SAR_p) and the electrolyte content of the solution, which can be expressed by either the electrical conductivity (EC_{25}) or the sum of the cations ($\sum cat$) in the fluid.

Chemical conditions also influence another dynamic property of cohesive materials relevant to rill initiation, the ability to shrink and swell. In badland regolith materials containing smectite clays this can produce a highly porous and permeable structure which permits rapid internal drainage (IMESON 1986). In extreme cases a surface "popcorn"-like layer characterized by an extremely low bulk density can develop (BRYAN et al. 1978). The macroporosity produced by shrinking and swelling in badland regolith materials may be analogous in its effects on internal drainage and rill initiation to the macroporosity produced in cultivated soils by tillage.

Ped slaking and surface crusts can also contribute to rill development. DE PLOEY & MUCHER (1981) developed the C_{5-10} index from Atterberg consistency limits to indicate the susceptibility of Belgian loess soils to crust formation. Consistency indices indicate the capacity of material to withstand flow or plastic deformation in terms of its moisture content. On badland slopes, low liquid or plastic limits could be expected to encourage mass movement which inhibits or obliterates rill development.

1.2 REGOLITH ON BADLANDS SLOPES

Regolith profiles on badland slopes have been described by several authors (BRYAN et al. 1978, 1984, YAIR et al. 1980, HODGES 1984, GERITS 1986). Rather similar profiles are found on mudstones in the Dinosaur Park badlands, Alberta, and marls in southeastern Spain, despite significant differences in parent material mineralogy. The depth of the regolith profile (typically 30–40 cm) is believed to represent the typical depth of water penetration, with localised deeper penetration along major cracks, tunnels and pipes. At most sites the following profile sequence occurs:

1. **crust, 1–2 cm thickness** Surface crust broken by desiccation cracks which evolve through time. May take various forms depending on age, exposure, subtle variations in clay content and the presence of blue-green algae (FINLAYSON & GERITS 1986).

2. **subcrust, $\pm$ 10 cm** Layer with high macroporosity when dry, but compact when wet. Highly dynamic and often characterised by micropipes.

3. **weathering layer, $\pm$ 15 cm** Transitional layer of porosity with partially weathered shards clearly recognizable.

4. **shard layer, $\pm$ 20–40 cm** Subangular fragments or shards into which parent rock breaks on wetting or weathering.

5. **parent rock.**

The properties of typical profiles representative of major lithological units are shown in tab.3 and 4.

The profiles described appear to be rather typical of bedland slopes developed on smectite clays, which include the areas discussed in this paper, and the Zin badlands in Israel, described by YAIR et al. (1980). Most profiles contain horizons of high macroporosity and permeability, which results in migration and growth of salt crystals. Sometimes a very fine granular layer with large amounts of sodium chloride, sodium sulphate or calcium sulphate is present. Profiles may vary considerably on homogeneous lithology as a result of aspect differences and their influence on moisture balance and weathering (YAIR et al. 1980, FINLAYSON & GERITS 1986).

Although the profile described above is typical of many badland slopes, some lithological units show extremely limited profile development. These include the muddy siltstones and fine sandstones of the Dinosaur Park badlands which

Regolith material	% clay $\bar{x}$	s	% silt $\bar{x}$	s	% sand $\bar{x}$	s	% CaCO$_3$ $\bar{x}$	s	% CaSO$_4$.2H$_2$O $\bar{x}$	s
Dehesas										
crust	24.4	4.4	65.9	5.5	9.7	3.2	55.0	7.3	3.0	0.7
subcrust	27.8	6.6	63.8	4.4	8.4	4.3	52.8	5.4	2.9	0.4
weath. l.	30.4	6.4	61.3	5.1	8.3	4.7	50.5	7.3	3.2	1.0
Cortijo										
crust	26.4	4.8	63.7	3.2	10.7	5.2	61.7	8.2	2.4	0.9
subcrust	30.2	6.3	60.5	3.4	9.3	6.1	61.5	8.0	2.2	0.8
weath. l.	27.9	4.2	62.9	4.1	9.2	4.6	59.1	6.0	2.5	0.8

Table 3: *Profile characteristics (weight % of fine earth) of regolith from badlands in Spain ($\bar{x}$ = mean value, s = standard deviation).*

Microcatchment	Regolith material		% Clay	% Silt	% Sand
1	grey shale		55.5	36.5	8.0
	surface		62.0	32.0	6.0
	crust		-	-	-
	shard		65.0	33.0	2.0
	yellow shale		45.5	49.0	5.6
	surface		57.0	35.0	8.0
	crust		61.0	35.0	4.0
	shard		49.0	49.0	2.0
	sandstone		22.0	33.0	45.0
			35.0	38.0	27.0
	pediment		9.0	54.5	35.7
			10.0	62.0	28.0
2	grey shale	(unit 1)			
	surface		61.0	33.0	6.0
	crust		66.0	30.0	4.0
	shard		51.0	39.0	10.0
	grey shale	(unit 2)			
	surface		62.0	35.0	3.0
	crust		57.0	38.0	5.0
	shard		96.0	4.0	0.0
	grey shale	(unit 3)			
	surface		65.0	32.0	3.0
	crust		64.0	33.0	3.0
	shard		59.0	37.0	4.0
3	yellow shale		56.5	40.0	4.5
	surface	0–2 cm	64.5	31.5	3.4
		2–4 cm	73.5	24.5	1.6
	crust	4–8 cm	91.5	7.0	1.6
	shard	> 10 cm	68.0	29.0	3.3
4	grey shale	(unit 1)			
	surface		58.0	38.0	4.0
	crust		-	-	-
	shard		63.0	34.0	3.0
	grey shale	(unit 2)			
	surface		69.0	27.0	4.0
	crust		-	-	-
	shard		63.0	31.0	6.0

Table 4: *Profile characteristics of regolith from Dinosaur Provincial Park badlands.*

are almost impermeable and develop a very thin weathering rind (HODGES & BRYAN 1982). Rill development on these units is, nevertheless, pronounced (BOWYER-BOWER & BRYAN 1986).

2 FIELD SITES

The Spanish field sites are located at Dehesas de Guadix and Cortijo in the valleys of the Rio Fardes and Rio Guadahortuna in the Province of Granada (fig.1), in a semi-arid area with mean annual pricipitation of 300 mm, most of which falls between October and May. Both areas are developed in marine marls and average amplitude of relief at both sites is 30 m. Those at Dehease (phot.1) are of Miocene age and contain evaporite minerals like gypsum and halite. At Cortijo the marls are of Eocene age, and regolith composition superficially resembles that at Deheses (tab.1 and 3). However, for related regolith types with comparable salinity, the SAR_p values at Cortijo are higher, and sodium sulphate salts are relatively more important, as shown by the sodium/chloride and calcium/sulphate ratios (tab.1). The Cortijo site is characterized by dense networks of subparallel rills (phot.2), individual channels usually being 5–15 cm wide and 2–10 cm deep. At Dehesas, on the contrary, rills are very infrequent. Shallow subsurface drainage networks of micropipes which feed into larger pipes or gullies occur at both sites.

The Dinosaur Park badlands (fig.2) have been described by CAMPBELL (1970, 1977), BRYAN et al. (1978) and HODGES & BRYAN (1982). They have developed in bentonitic mudstones (earlier described as shales), siltsones and fine sandstones of the Upper Cretaceous Judith Formation (KOSTER 1983). The climate is semi-arid with a mean annual precipitation of 340 mm. Seventy per cent of the precipitation falls between April and September in a mixture of low intensity frontal and moderate-high intensity convective rainfall. The remainder occurs as snow during the intensely cold and prolonged winter. Badland development is more extensive and more complex than at the Spanish sites, and is incised to 60 m below the surrounding prairie surface. HODGES & BRYAN (1982) have described the most important lithology units as grey mudstone, yellow mudstone and sandstone (phot.3 and 4). Different units are often thin, and completely interbedded, so that rill systems are much less homogeneous than at the Cortijo.

Characteristic regolith properties for mudrocks in the Dinosaur Park badlands are shown in tab.4. Regolith is shallower than in the Spanish sites, and contains lower amounts of water soluble salts (tab.2). As at Cortijo, however, sodium salts are important, and clays are highly dispersible. The mudstones can have clay contents exceeding 90%, most of which are <1 micron in diameter, and have extremely high swelling capacities (tab.8). When dehydrated mass bulk-densities are very low (0.4 to 1.1 g cm^{-3}).

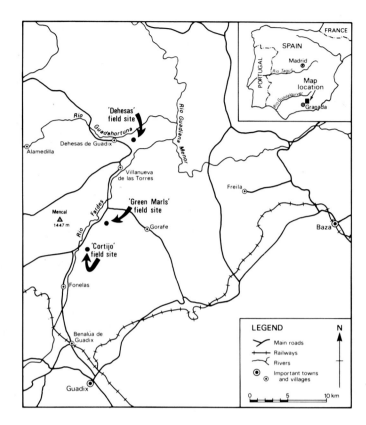

Figure 1: *Location of field sites in southeastern Spain.*

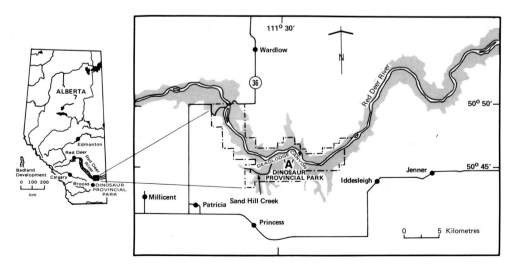

Figure 2: *Location of field site in southeastern Alberta.*

Photo 1: *Badland slopes at the Dehesas site. Regolith material is transported to channels by shallow mass-wasting..*

3 METHODS

The analyses reported in this paper were carried out as part of an investigation of erosional processes in badland areas in Alberta (BRYAN et al. 1978, 1984, HODGES & BRYAN 1982) and by GERITS (1986) in southeastern Spain. In previous papers details of the general objectives, the location and selection of sampling sites and the procedures used to determine physical, mineralogical and chemical properties of materials have been given. Only the methods used to characterize the response of regolith horizons to wetting, the consistency, shrinking and swelling behaviour, and dispersion characteristics determined from samples for which chemical and mineralogical data are available, are described here.

Consistency parameters were determined using standard methods following the procedures given in SINGH (1967). The plastic (W_p) and liquid (Wl) limits were used to calculate the plasticity index (I_p) and the activity (A_i) (YOUNG & WARKENTIN 1975). The C_{5-10} index (DE PLOEY & MÜCHER 1981) was calculated from the characteristics of the liquid limit curve.

Shrink-swell parametrs were determined in several ways. For samples from Spain and some samples from Canada large clods of regolith fragments were allowed to swell at pF 1, until they attained constant weight. These were then coated in Saran^R resin and allowed to

Photo 2: *Badland slopes at the Cortijo site. Regolith material is dissected by a dense rill network, and is transported to channels by rillwash.*

Photo 3: *Rill development on smectitic mudstones at Dinosaur Provincial Park, Alberta.*

Photo 4: *Rill development on fine-grained sandstones at Dinosaur Provincial Park, Alberta.*

Photo 5: *Grey mudstone site at Dinosaur Park showing conspicuous mound and depression microtopography.*

dry slowly for several weeks, while periodic volume and mass measurements were made. From these data COLE factors at different moisture contents were calculated and plotted against gravimetric moisture content. With this curve ("COLE" = $f(\theta_g)$) the following parameters were estimated: shrinkage limit (W_s), slope of the linear shrinkage line (S_l), COLE factor at 25% gravimetric moisture ($COLE_{25}$), and the dry bulk density (BD_d). The samples from Canada could not be treated in this way because they became too soft and fluid when allowed to swell freely to permit treatment with resin. Volumetric properties were determined from sequential photographs taken as samples were moistened and dried (IMESON 1986), and by determination of the amount of water the samples could absorb when allowed to swell freely.

The dispersion index (D.I.) used is a modification of Middleton's dispersion ratio (MIDDLETON 1930), using 0.002 mm instead of 0.005 mm as an arbitrary value to determine the amount of dispersible clay.

4 CONSISTENCY, VOLUMETRIC AND DISPERSIVE CHARACTERISTICS OF REGOLITH SAMPLES

4.1 CONSISTENCY

Average consistency parameter values for the Dehesas and Cortijo profiles are summarised in tab.5. The same trends can be seen at both locations. The C_{5-10} index is relatively low for the crust samples, particularly those from Dehesas, and all other properties increase in value with depth, although the increase for the plastic limit is small. When equivalent horizons are compared, those from Cortijo show higher activity, liquid limit, plasticity index and C_{5-10} index values. These differences are related to the $CaCO_3$ content, clay mineralogy and chemical composition (particularly the dominance of Na) of the regolith material. Higher liquid limit, activity and C_{5-10} index values at Cortijo show that this material will not become fluid or unstable at the same moisture content threshold as the Dehesas material.

Data from Canada (tab.6) show much higher liquid limits and plasticities, and lower plastic limits than for the Spanish materials, due to differences in clay content and mineralogy. This would explain the infrequency of fluid mass movement processes in the Dinosaur Park badlands, while slow regolith creep and "dry" aggregate falls due to instability triggered by restricted surface swelling, are comparatively common (HODGES & BRYAN 1982). Variation in properties between different mudstone units is also higher than at the Spanish sites.

4.2 VOLUMETRIC PROPERTIES

The relative importance of volumetric properties at Dehesas and Cortijo can be assessed from $COLE_{25}$ and S_l values (tab.7). Material from Cortijo has far greater capacity for volumetric change, with $COLE_{25}$ values 60% greater than those of the equivalent Dehesas horizons. The lower shrinkage limit for the Cortijo samples suggests that this material begins to swell at lower moisture levels. Dry bulk densities (BD_d) of natural clods from Cortijo are also higher. Because of the size of the clods used ($\pm$ 30 cm^3), it was not possible to estimate the macroporosity of the subcrust samples. The true dry bulk density of the subcrust horizons is probably considerably lower than the values shown in tab.7.

The Dinosaur Park samples are subject to large volumetric changes for the reasons mentioned. The amount of water absorbed at tension pF 1 indicates the amount of water which undisturbed shards or peds can take up (tab.8). This is usually an order of magnitude greater than the shrinkage limit (SL) and far higher than comparable values obtained for the Spanish samples. The value of the volumetric shrinkage (VS, tab.8) refers to the moisture contents of saturated pastes for determination of SL and SR (SINGH, 1967), which values are far higher than Sw (pF 1) values, but generally show similar trends. The grey mudstone samples were collected from a relatively level site without rill development, but which showed a conspicuous microtopography of mounds and

Regolith layer	average clay content %	C_{5-10} index		plastic limit Wp		liquid limit WL		plasticity index Ip		Activity Ai	
		$\bar{x}$	s	$\bar{x}$	s	$\bar{x}$	s	$\bar{x}$	s	$\bar{x}$	s
Dehesas											
crust	24	2.70	0.81	22.20	3.3	35.4	4.3	13.2	4.2	0.55	0.20
subcrust	28	4.30	1.02	21.60	1.8	38.3	5.5	16.7	5.0	0.57	0.11
weath. l.	30	4.30	1.17	22.60	1.8	41.3	7.6	18.8	7.4	0.60	0.22
Cortijo											
crust	26	3.48	0.71	20.36	1.9	39.6	5.4	19.2	4.1	0.73	0.11
subcrust	30	4.93	1.36	21.10	1.7	46.2	8.1	25.1	6.8	0.80	0.11
weath. l.	28	4.50	1.26	21.60	1.6	47.3	3.0	25.7	2.8	0.89	0.15

Table 5: *Consistency parameters for regolith profiles at the Dehesas and Cortijo locations.*

Microcatchment	Regolith	Wp	Wl	Jp
1	Grey Shale 1	12.6	89.5	76.9
	Grey Shale 2	12.6	78.0	65.4
	Grey Shale 3	19.1	99.0	79.9
2	Grey Shale	7.0	93.5	86.5
	Yellow Shale	27.1	89.5	62.4
	Sandstone	8.6	84.5	75.9
	Pediment	21.4	27.0	5.6
4	Grey Shale 1	19.5	63.0	43.5
	Grey Shale 2	22.0	107.0	85.0

Table 6: *Consistency parameters for regolith profiles at Dinosaur Park (applied to surface units in tab.4).*

Regolith layer	% clay		BD_d		Shrinkage limit (Ws)		Coefficient of linear shrinkage $(S_l)^*$		$COLE_{25}$	
	$\bar{x}$	s	$\bar{x}$	s	$\bar{x}$	s	$\bar{x}$	s	$\bar{x}$	s
Dehesas										
crust	24.4	4.36	1.57	0.21	12.0	6.40	2.25	0.78	0.038	0.008
subcrust	27.8	6.60	1.63	0.14	14.5	4.50	2.95	0.89	0.046	0.013
waeth. l.	30.4	6.30	1.70	0.08	n.d.	n.d.	3.50	0.80	0.050	0.009
Cortijo										
crust	26.4	4.80	1.78	0.13	11.1	2.67	3.12	0.61	0.059	0.012
subcrust	30.2	6.20	1.79	0.018	10.6	1.90	3.82	0.52	0.069	0.009
waeth. l.	27.8	4.20	1.90	0.03	9.3	1.15	4.20	0.40	0.080	0.005

* Slope of the linear shrinkage curve (Cole vs % moisture)

Table 7: *Volumetric parameters for regolith profiles at the Dehesas and Cortijo locations.*

depressions (phot.5). Crust samples from depression sites showed higher silt concentrations, reflecting very localized colluviation (BRYAN et al. 1978), while the samples from mounds had a greater propensity for swelling. Unfortunately, no data for rilled grey mudstone sites are available. Yellow mudstone samples were collected from both rilled and unrilled sites, but at neither did mound and depression microtopography appear. Samples from the rilled location showed higher shrink-swell capacity (tab.8).

		% H$_2$O		volumetric shrinkage (VS)	average moisture content (Sw pF 1)
		shrinkage rates (SR)	shrinkage limit (SL)		
Grey shale (level sites, no rills)					
crust	mound	1.95	15.4	447	214
	depression	1.95	23.8	226	154
subcrust	mound	1.90	22.7	531	164
	depression	1.97	12.0	461	160
weath. l.		2.00	08.9	629	193
shards		1.86	18.8	507	095
crust	mound	1.89	14.9	263	066
	depression	2.13	13.1	232	072
subcrust		1.89	14.1	392	067
weath. l.		2.05	12.4	389	060
shards		1.84	14.0	377	150
Yellow shale (no rills)					
crust		1.94	12.6	295	050
subcrust		1.95	13.8	231	079
weath. l.		1.99	15.3	301	093
Yellow shale (rilled)					
crust		1.80	22.3	316	117
subcrust		2.00	16.8	387	124
weath. l.		2.00	16.6	342	122
shards		1.85	21.9	318	133

Table 8: *Volumetric parameters for regolith profiles at Dinosaur Park.*

4.3 DISPERSIVE CHARACTERISTICS

The parameters which indicate dispersive capacity at Dehesas and Cortijo are very variable, due to local site conditions (tab.9a). This can be reduced by stratifying the samples according to crust type (FINLAYSON & GERITS 1986). In spite of the variability the general trends are clear. The dispersion index (DI) and SAR$_p$ values are highest in the subcrust, while salt content increases with depth. The subcrust and weathering layers have comparable SAR$_p$ values, but quite different salinities, while the reverse holds for the crust and subcrust. The subcrust therefore has the most favourable conditions for dispersion. Since the Cortijo regolith is relatively less saline and has higher SAR$_p$ values, dispersive conditions will be reached more rapidly during rainfall than at Dehesas.

Only limited information is available about the dispersive characteristics of samples from Dinosaur Park (tab.9b). The dispersive capacity of surface horizons is much higher than at either Dehesas or Cortijo, with values very similar to the subcrust at Dehesas.

It can be concluded that regolith horizons at Cortijo are more dispersible, more liable to slaking and have a much higher potential for volumetric change than those at Dehesas, while Dinosaur Park materials, and particularly from surface horizons, generally display even higher activity. The relationship between these differences and rill development in both areas is discussed below.

Regolith layer	dispersion index (DI)		SAR$_p$* (mmol/l)$^{1/2}$		EC$_{25}$* (mS/cm)	
	$\bar{x}$	s	$\bar{x}$	s	$\bar{x}$	s
Dehesas						
crust	26.5	13.8	15.6	14.0	3.67	3.49
subcrust	44.6	27.2	26.4	18.6	6.05	6.07
weath. l.	22.2	17.9	25.9	15.9	11.14	8.22
Cortijo						
crust	32.4	21.5	15.9	09.5	3.05	2.79
subcrust	51.7	31.2	25.1	13.6	3.90	2.70
weath. l.	35.9	33.5	22.7	12.3	7.21	4.67

* (of 1:2 soil-water extracts)

Table 9: *a) Parameters related to dispersion at the Dehesas and Cortijo locations.*

	dispersion index* DI (%)		SAR$_p$* (mmol/l)$^{1/2}$	EC$_{25}$ (mS/cm)
	$\bar{x}$	s	$\bar{x}$	$\bar{x}$
Upper grey shale				
surface	42.9	61.6	32.1	4.74
subsurface	09.9	09.1	-	-
Sandstone				
surface	04.8	02.0	22.0	2.27

* saturated paste extracts

Table 9: *b) Parameters related to dispersion at the Dinosaur Park badlands.*

5 DISCUSSION

5.1 RILLS AS DRAINS ON BADLAND SLOPES

It is possible to understand how the properties discussed above affect rill development on badland slopes through their influence on regolith hydrology by considering rills, micropipes and cracks as miniature field drains. Drains are designed to control water table depth and various equations have been developed to determine the appropriate drain spacing for given conditions (WESSELING 1973). An example is a steady state drainage equation describing horizontal flow to ditches reaching an impervious floor, derived from the Donnan equation. It considers a two-layered soil with interface at drain level and describes subsurface flow above and below the drain:

$$Q = \frac{(8.Kb.D.h) + (4.Ka.h^2)}{L^2} \quad (1)$$

where
Q = drain discharge,
K = hydraulic conductivity above (Ka) or below (Kb) drain level,
D = thickness of aquifer below drain level,
h = hydraulic head for subsurface flow into drains,
L = drain spacing.

(WESSELING 1973).

If rills, micropipes and cracks are considered as drains in (partly) saturated regolith, this equation can be applied for different drain levels at successive stage during runoff (fig.3). The first term of the equation can, in fact, be ignored, since D can be seen as zero.

True saturation of smectite regolith can be achieved only rather slowly, but crust and subcrust materials on badland slopes become effectively saturated rather quickly during rainfall as wetting and swelling causes the surfaces of peds to seal, and some cracks to close. Water from the crust and subcrust drains laterally into subsurface cracks. The rate of drainage depends on the hydraulic conductivity of the weathering layer, the macroporosity of the subcrust, the hydraulic gradient (h) and the crack density (L). Low crust and subcrust hydraulic conductivity, deep cracks and a high crack density cause a steep hydraulic gradient towards subsurface cracks. The macropores of the highly dispersible subcrust form micropipes which also drain laterally into these cracks.

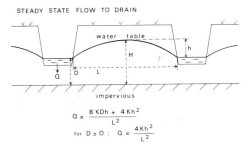

Figure 3: *Schematic representation of drainage model.*

The results of these processes is the development of a dense subsurface network of micropipes and cracks that produces preferential flow paths of water both on and beneath the surface. Areas close to major cracks become preferential drainage locations, where erosion of soft and swollen material leads to enlargement of cracks into micropipes, or, by collapse, into rills. Intervening areas between major cracks are less well-drained, and tend to remain as residual mounds or interfluves.

In theory, an evolutionary sequence of rill development could be envisaged, in which a slope is initially drained by a micropipe system, which in turn gives way to a mixture of micropipes and rills, in which surface hydraulic conditions become progressively more dominant, particularly as local microtopographic evolution and colluviation lead to increasing differentiation of runoff response. In practice all stages usually coexist on the same hillslope, so that hillslope runoff typically includes surface and subsurface components, and it is not possible to make a hydrologic distinction between rill and micropipe flow (BRYAN et al. 1978).

Following the "drainage theory", micropipe formation depends on the hydraulic properties of the material and crack spacing. In fact, cracking in smectitic regolith is so dense and apparently so homogeneous, that the preferential development of certain cracks into micropipes depends primarily on material hydraulic properties. If the regolith is essentially homogeneous, these properties will depend on water concentration, which controls the rate and extent of swelling. A minimal threshold three-dimensional drainage area will be necessary to concentrate sufficient water for micropipe evolution, which will vary in size with the regolith characteristics. Ultimately this will determine the spacing of rills, and the distance below an interfluve at which they appear. This leads to a concept of threshold drainage area for micropipe development and rill initiation analogous to the "constant of channel maintenance" which SCHUMM (1956) introduced for surface drainage networks in the Perth Amboy badlands.

Once micropipes collapse to form rills, subsequent development depends on the hydraulic conditions of surface flow (shear stress and velocity) and on surface material characteristics (critical shear stress, cohesion, liquefaction, roughness), all of which are strongly influenced by changes in moisture content. As noted earlier, rill patterns can vary, even in areas of uniform lithology, due to localised differences in moisture balance. Even when rills become dominant, the "drainage theory" remains applicable, as the rills now function as surface drains.

5.2 REGOLITH BEHAVIOUR UNDER SIMULATED RAINFALL

The hypothesis that rills, micropipes and cracks act as drains is supported by simulated rainfall experiments in badland areas in southeastern Spain and western Canada.

GERITS (1987) carried out 10 simulated rainfall experiments at Dehesas and Cortijo, in Spain, on hillslope microcatchments which ranged from 50–100 m^2 in area. Rainfall was applied at intensities of 25–50 mm.h.$^{-1}$ for 40–60 minutes. The sites showed comparable rates of runoff, but sediment concentrations at Cortijo were almost an order of magnitude higher, and runoff also had a much higher SAR$_p$ value.

Visual observation of dyes showed that runoff at Cortijo moved only short distances over the surface (50–100 cm) before draining into rills and gullies. Rills are connected and fed by a complex of micropipes and subsurface cracks which drain the regolith. Seepage of pipe and crack flow on steep slopes and gully walls were frequent, resulting in local mudflows. At Dehesas runoff occurred mainly as overland flow draining into larger gullies, and extensive shallow mass movements also occurred. Pipe and crack flow were of minor importance, despite the presence of macropores in a dispersible subcrust. In contrast with Cortijo there were few clay coatings on the walls of cracks and macropores as an index of significant subsurface discharge.

Rill development is affected by both the vertical and horizontal distribution of water in the regolith. Although the Cortijo regolith has a higher moisture retention capacity, profile moistening during rainfall is shallower than at Dehesas, due to lower crust infiltration capacities resulting from swelling. Together with the occurrence of more frequent and deeper cracks, this leads to more favourable conditions for the development of a steep hydraulic gradient at Cortijo. The Cortijo subcrust with macropores is more dispersible and micropipes appear to develop more frequently. Swelling may be a critical control on the development of subsurface flow, and if it is too extensive, may seal the surface completely, so that drainage into the subcrust is insufficient to generate subsurface flow.

The Cortijo surface materials show considerable spatial variation in drainage, which results in a parallel variation in physico-chemical processes, consistency, shear strength and critical shear stress. Where drainage is impeded, high moisture contents and flow concentration encourage swelling, dispersion, liquefaction and solute release, which in turn affects the roughness of the surface, and therefore flow hydraulics (GERITS 1987).

The field experiments showed that swelling and dispersion are dominant at Cortijo, while flocculation is more important at Dehesas. Simulated rainfall tests with fluorescent dyes showed that spatial variation in the moisture content of surface materials was much lower at Dehesas than Cortijo, and so the processes of moisture concentration and rill development are less intense.

Numerous simulated rainfall tests have been carried out in Dinosaur Park on hillslope microcatchments ranging in area from 10 m²–45 m² in area (BRYAN et al. 1978, HODGES 1982, BRYAN & HODGES 1984, BRYAN et al. 1984). There were some variations in procedures but most tests were at a rainfall intensity of 29 mm.h.$^{-1}$ for 30–45 minutes. Lithology variations amongst smectitic mudstones in Dinosaur Park are much greater than at Dehesas or Cortijo, and further complexity is added by interbedding of fine and coarse sandstones, siltstones, siderite layers, carbonaceous shales and depositional micropediments. Surface and subsurface crack patterns are affected by spatial and temporal variations in physical and chemical properties, and by differences in the incidence and rate of wetting-drying and freeze-thaw cycles. Despite these extremely complex variations, development of surface drainage networks on all mudstones appears to be preceded by subsurface micropipe development, and rill formation therefore conforms to the "drainage theory". Subsurface flow concentration originating in overlying mudstone layers is an essential precursor of rill initiation on fine smectitic sandstones and siltstones (BOWYER-BOWER & BRYAN 1986), while rill chute development on micropediments can be explained by the hydraulic characteristics of runoff alone (HODGES 1982). All the mudstones tested behaved more like the Cortijo materials than like those at Dehesas, with micropipe and rill development rather than mass movement. At one steeply sloping site, however, two tests in succeeding years produced a mudflow in the first test (BRYAN et al. 1978) and rill development in the second (HODGES 1984). It is believed that this result was caused by changes in the hydraulic conductivity of the subcrust reflecting alterations in regolith chemistry. This serves to emphasize the very complex evolutionary sequence which rill development can follow on smectitic mudstones on badland slopes.

6 CONCLUSIONS

By considering rills as drains it is possible to explain why some rill systems occur in association with particular regolith properties. The observations reported here show that rills occur on smectitic materials which respond in a particular way to wetting. They are most likely to occur on materials with a high shrink-swell capacity, which are subject to dispersion and which slake to form crusts. These responses influence the hydraulic and hydrologic characteristics in such a way that drainage occurs discontinuously through macropores and cracks. These cracks act as drains which concentrate flow and cause localised saturation. This is important because swelling and dispersion can lead to high rates of erosion as saturated regolith can flow spontaneously in miniature mudflows, and because the saturated dispersible material has little resistance to entrainment.

The essential conditions for rill development on these materials are the occurrence of discontinuities in the flow and the presence of interconnected macropores or cracks. Discontinuities are favoured by hysteresis in shrinking and swelling which ensures that some cracks usually remain open as the regolith is moistened. Swelling and the sealing of ped surfaces ensures that cracks can develop as drains, ultimately forming rills.

It is interesting to consider the implications for rill development on cultivated soils. Where ploughing produces peds which can seal by slaking, and where underlying horizons are not very permeable, conditions suitable for rill development, analogous to those on badland slopes, could be created. Rill development would require a ploughed soil horizon with high macroporosity, susceptibility to slaking and swelling and, under some conditions, a relatively high liquid limit. Whether or not these properties can be used to identify areas of rill erosion hazards remains to be investigated.

ACKNOWLEDGEMENT

The research reported in this paper was supported by the Netherlands Foundation for Pure Scientific Research (ZWO), the University of Amsterdam and the Natural Sciences and Engineering Council, Canada.

REFERENCES

[ABRAHAMS 1968]] ABRAHAMS, A.D.: Distinguishing between the concepts of steady state and dynamic equilibrium in geomorphology. Science, J. **2**, 160–166.

[ARIATHURAI 1978] ARIATHURAI, K.: Erosion rates of cohesive soils. Journal of the Hydraulics Division, ASCE, **104**, HY 2, 279–284.

[ARULANDAN et al. 1975] ARULANDAN, K., LOGANATHAN, P. & KRONE, R.B.: Pore and eroding fluid influences on surface erosion fo soil. Journal of the Geotechnical Engineering Division, ASCE, **101**, GT 1, 51–66.

[BOWYER-BOWER & BRYAN 1986] BOWYER-BOWER, T.A.S. & BRYAN, R.B.: Rill initiation: concepts and experimental evaluation on badland slopes. Zeitschrift für Geomorphologie, Suppl. Bd. **59**, 161–175.

[BRYAN et al. 1978] BRYAN, R.B., YAIR, A. & HODGES, W.K.: Factors controlling the initiation of runoff and piping in Dinosaur Provincial Park badlands, Alberta, Canada. Zeitschrift für Geomorphologie, Suppl. Bd. **29**, 151–168.

[BRYAN & HODGES 1984] BRYAN, R.B. & HODGES, W.K.: Runoff and sediment transport dynamics in Canadian badland microcatchments. In: Catchment Experiments in Geomorphology. (Ed. D.E. WALLING & T.P. BURT), GeoBooks, Norwich, 115–132.

[BRYAN et al. 1984] BRYAN, R.B., IMESON, A.C. & CAMPBELL, I.A.: Solute release and sediment entrainment on microcatchments in the Dinosaur Badlands, Alberta, Canada. Journal of Hydrology, **71**, 79–106.

[BRYAN & CAMPBELL 1986] BRYAN, R.B. & CAMPBELL, I.A.: Runoff generation and sediment transport in a semi-arid ephemeral drainage, basin. Zeitschrift für Geomorphologie, Suppl. Bd. **58**, 121–143.

[CAMPBELL 1970] CAMPBELL, I.A.: Erosion rates in the Steveville badlands, Alberta, Canada. Canadian Geographer, **14**, 202–216.

[CAMPBELL 1977] CAMPBELL, I.A.: Stream discharge, suspended sediment and erosion rates in the Red Deer River Basin, Alberta, Canada. International Association of Hydrological Sciences, Publication **122**, 244–259.

[DE PLOEY & MÜCHER 1981] DE PLOEY, J. & MÜCHER, H.J.: A consistency index and rainwash mechanisms on Belgian loamy soils. Earth Surface Processes, **6**, 319–330.

[DE PLOEY 1984] DE PLOEY, J.: Hydraulics and runoff and loess deposition. Earth Surface Processes and Landforms, **9**, 533–540.

[FINLAYSON & GERITS 1986] FINLAYSON, B.L. & GERITS, J.: Crusted microtopography on badland slopes in SE Spain, CATENA (in press.)

[GERITS 1986] GERITS, J.J.P.: Regolith properties and badland development. pp. 71–74 in: Estudios sobre Geomorfologia del Sur de Espana. (Ed. F. LOPEZ BERMUDEZ & J.B. THORNES), Murcia.

[GERITS et al. 1986] GERITS, J.J.P., IMESON, A.C. & VERSTRATEN, J.M.: Chemical thresholds and erosion in saline and sodic materials. pp. 75–79 in: Estudios sobre Geomorfologia del Sur de Espana. (Ed. F. LOPEZ BERMUDET & J.B. THORNES), Murcia.

[GERITS 1986] GERITS, J.J.P.: Implications of chemical threshold and physico-chemical processes for modelling erosion in S.E. Spain. Paper presented to the COMTAG Symposium, Granada, September 14, 1986.

[GERITS 1987] GERITS, J.J.P.: Physico-chemical thresholds for sediment entrainment and transport. University of Amsterdam, Ph.D., in preparation.

[GOVERS 1985] GOVERS, G.: Selectivity and transport capacity of thin flow in relation to rill erosion. CATENA, **12**, 35–49.

[GRISSINGER 1966] GRISSINGER, E.H.: Resistance of selected clay systems to erosion by water. Water Resources Research, **2**, 131–138.

[HODGES 1982] HODGES, W.K.: Hydraulic characteristics of a badland pseudo-pediment slope system during simulated rainfall experiments. In: Badland Geomorphology and Piping. (Ed. R.B. BRYAN & A. YAIR), GeoBooks, Norwich, 127–152.

[HODGES 1984] HODGES, W.K.: Experimental study of hydrogeomorphological processes in Dinosaur Badlands, Alberta, Canada. University of Toronto, Ph.D. thesis.

[HODGES & BRYAN 1982] HODGES, W.K. & BRYAN, R.B.: The influence of material behaviour on runoff initiation in the Dinosaur Badlands, Canada. In: Badland Geomorphology and Piping. (Ed. R.B. BRYAN & A. YAIR), GeoBooks, Norwich, 13–46.

[HORTON 1945] HORTON, R.B.: Erosional development of streamsand their drainage basins: hydrophysical approach to quantitative morphology. Bulletin Geological Society of America, **56**, 275–370.

[IMESON 1986] IMESON, A.C.: Investigating volumetric changes in clayey soils related to subsurface water movement and piping. Zeitschrift für Geomorphologie, Suppl. Bd. **59**, 115–130.

[IMESON & VERSTRATEN 1986] IMESON, A.C. & VERSTRATEN, J.M.: Erosional thresholds related to the chemical and physical processes of badland regolith materials in SE Spain. Paper presented to COMTAG Symposium, Granada, September 14, 1986.

[KOSTER 1983] KOSTER, E.H.: Sedimentology of the Upper Cretaceous Judith River (Belly River) Formation, Dinosaur Provincial Park, Alberta. Alberta Geological Survey, Edmonton.

[LEOPOLD & LANGBEIN 1962] LEOPOLD, L.B. & LANGBEIN, W.B.: The concept of entropy in landscape evolution. United States Geological Survey Professional Paper 500A.

[MEYER & MONKE 1965] MEYER, L.D. & MONKE, E.J.: Mechanics of soil erosion by rainfall and overland flow. Transactions of the American Society of Agricultural Engineers, 572–577, 580.

[MIDDLETON 1930] MIDDLETON, H.E.: Properties of soils which influence soil erosion. United States Department of Agriculture Technical Bulletin, 178.

[PARTHENIADES 1972] PARTHENIADES, E.: Results of recent investigations on erosion and deposition of cohesive sediments. In: Sedimentation. (Ed. H.W. SHEN), Fort Collins, Colorado.

[SARGUNAM et al. 1973] SARGUNAM, A., RILEY, P., ARULANANDAN, K. & KRONE, R.B.: Physico-chemical factors in erosion of cohesive soils. Journal of the Hydraulics Division, ASCE 99, HY 3, 555–558.

[SAVAT 1979] SAVAT, J.: Laboratory experiments on erosion and sedimentation of loess by laminar sheer flow and turbulent rill flow. In: Erosion Agricole des Sols en Milieu Tempéré Non-Mediterrane. (Ed. H. VOGT & T. VOGT), Strassbourg, 139–143.

[SAVAT & DE PLOEY 1982] SAVAT, J. & DE PLOEY, J.: Sheetwash and rill development by surface flow. In: Badland Geomorphology and Piping. (Ed. R.B. BRYAN & A. YAIR), GeoBooks, Norwich, 113–126.

[SCHUMM 1956] SCHUMM, S.A.: Evolution of drainage systems and slopes in badlands at Perth Amboy, N.J. Bulletin of the Geological Society of America, **67**, 597–646.

[SCHUMM & LUSBY 1963] SCHUMM, S.A. & LUSBY, G.C.: Seasonal variations of infiltration capacity and runoff on hillslopes in western Colorado. Journal of Geophysical Research, **68**, 3655–3666.

[SINGH 1967] SINGH, A.: Soil Engineering in Theory and Practice. Asia Publishing House, London, 805 p.

[WESSELING 1973] WESSELING, J.: Drainage Principles and Applications, Vol. **2**. Theories of Field Drainage and Watershed Runoff, ILRI, Wageningen.

[YAIR et al. 1980] YAIR, A., BRYAN, R.B., LAVEE, H. & ADAR, E.: Runoff and erosion processes and rates in the Zin Valley badlands, Northern Negev, Israel. Earth Surface Processes, **5**, 205–226.

[**YONG & WARKENTIN 1975**] YONG, R.N. & WARKENTIN, B.P.: Introduction to Soil Behaviour. Elsevier, Amsterdam, 489 pp.

Adresses of authors:
J. Gerits, A.C. Imeson, J.M. Verstraten
Fysisch Geografisch en Bodemkundig Laboratorium, Universiteit van Amsterdam,
Dapperstraat 115, 1093 BS Amsterdam,
The Netherlands
R.B. Bryan
Department of Geography, University of Toronto
(Scarborough Campus),
1265, Military Trail,
Scarborough, Ontario L3P 3B5,
Canada

DAN H. YAALON (ED.)

ARIDIC SOILS and GEOMORPHIC PROCESSES

SELECTED PAPERS of the INTERNATIONAL CONFERENCE
of the INTERNATIONAL SOCIETY of SOIL SCIENCE
Jerusalem, Israel, March 29 – April 4, 1981

CATENA SUPPLEMENT 1, 1982

Price: DM 95,–

ISSN 0722–0723 / ISBN 3–923381–00–X

This CATENA SUPPLEMENT comprises 12 selected papers presented at the International Conference on Aridic Soils – Properties, Genesis and Management – held at Kiryat Anavim near Jerusalem, March 29 – April 4, 1981. The conference was sponsored by the Israel Society of Soil Science within the framework of activities of the International Society of Soil Science. Abstracts of papers and posters, and a tour guidebook which provides a review of the arid landscapes in Israel and a detailed record of its soil characteristics and properties (DAN et al. 1981) were published. Some 49 invited and contributed papers and 23 posters covering a wide range of subjects were presented at the conference sessions, followed by seven days of field excursions.

The present collection of 12 papers ranges from introductory general reviews to a number of detailed, process oriented, regional and local studies, related to the distribution of aridic soils and duricrusts in landscapes of three continents. It is followed by three papers on modelling and laboratory studies of geomorphic processes significant in aridic landscapes. It is rounded up by a methodological study of landform–vegetation relationships and a regional study of desertification. Additional papers, related to soil genesis in aridic regions, are being published in a special issue of the journal GEODERMA.

D.H. Yaalon
Editor

G.G.C. CLARIDGE & I.B. CAMPBELL
A COMPARISON BETWEEN HOT AND COLD DESERT SOILS AND SOIL PROCESSES

R.L. GUTHRIE
DISTRIBUTION OF GREAT GROUPS OF ARIDISOLS IN THE UNITED STATES

M.A. SUMMERFIELD
DISTRIBUTION, NATURE AND PROBABLE GENESIS OF SILCRETE IN ARID AND SEMI-ARID SOUTHERN AFRICA

W.D. BLÜMEL
CALCRETES IN NAMIBIA AND SE-SPAIN RELATIONS TO SUBSTRATUM, SOIL FORMATION AND GEOMORPHIC FACTORS

E.G. HALLSWORTH, J.A. BEATTIE & W.E. DARLEY
FORMATION OF SOILS IN AN ARIDIC ENVIRONMENT WESTERN NEW SOUTH WALES, AUSTRALIA

J. DAN & D.H. YAALON
AUTOMORPHIC SALINE SOILS IN ISRAEL

R. ZAIDENBERG, J. DAN & H. KOYUMDJISKY
THE INFLUENCE OF PARENT MATERIAL, RELIEF AND EXPOSURE ON SOIL FORMATION IN THE ARID REGION OF EASTERN SAMARIA

J. SAVAT
COMMON AND UNCOMMON SELECTIVITY IN THE PROCESS OF FLUID TRANSPORTATION:
FIELD OBSERVATIONS AND LABORATORY EXPERIMENTS ON BARE SURFACES

M. LOGIE
INFLUENCE OF ROUGHNESS ELEMENTS AND SOIL MOISTURE ON THE RESISTANCE OF SAND TO WIND EROSION

M.I. WHITNEY & J.F. SPLETTSTOESSER
VENTIFACTS AND THEIR FORMATION: DARWIN MOUNTAINS, ANTARCTICA

M.B. SATTERWHITE & J. EHLEN
LANDFORM–VEGETATION RELATIONSHIPS IN THE NORTHERN CHIHUAHUAN DESERT

H.K. BARTH
ACCELERATED EROSION OF FOSSIL DUNES IN THE GOURMA REGION (MALI) AS A MANIFESTATION OF DESERTIFICATION

F. Ahnert, H. Rohdenburg & A. Semmel:

BEITRÄGE ZUR GEOMORPHOLOGIE DER TROPEN (OSTAFRIKA, BRASILIEN, ZENTRAL- UND WESTAFRIKA) CONTRIBUTIONS TO TROPICAL GEOMORPHOLOGY

CATENA SUPPLEMENT 2, 1982
Price: DM 120,–
ISSN 0722–0723 / ISBN 3–923381–01–8

F. AHNERT
UNTERSUCHUNGEN ÜBER DAS MORPHOKLIMA UND DIE MORPHOLOGIE DES INSELBERGGEBIETES VON MACHAKOS, KENIA

(INVESTIGATIONS ON THE MORPHOCLIMATE AND ON THE MORPHOLOGY OF THE INSELBERG REGION OF MACHAKOS, KENIA)

S. 1–72

H. ROHDENBURG
GEOMORPHOLOGISCH–BODENSTRATIGRAPHISCHER VERGLEICH ZWISCHEN DEM NORDOSTBRASILIANISCHEN TROCKENGEBIET UND IMMERFEUCHT–TROPISCHEN GEBIETEN SÜDBRASILIENS

MIT AUSFÜHRUNGEN ZUM PROBLEMKREIS DER PEDIPLAIN–PEDIMENT–TERRASSENTREPPEN

S. 73–122

A. SEMMEL
CATENEN DER FEUCHTEN TROPEN UND FRAGEN IHRER GEOMORPHOLOGISCHEN DEUTUNG

S. 123–140

H.-R. BORK u. W. RICKEN

BODENEROSION, HOLOZAENE UND PLEISTOZAENE BODENENTWICKLUNG

SOIL EROSION, HOLOCENE AND PLEISTOCENE SOIL DEVELOPMENT

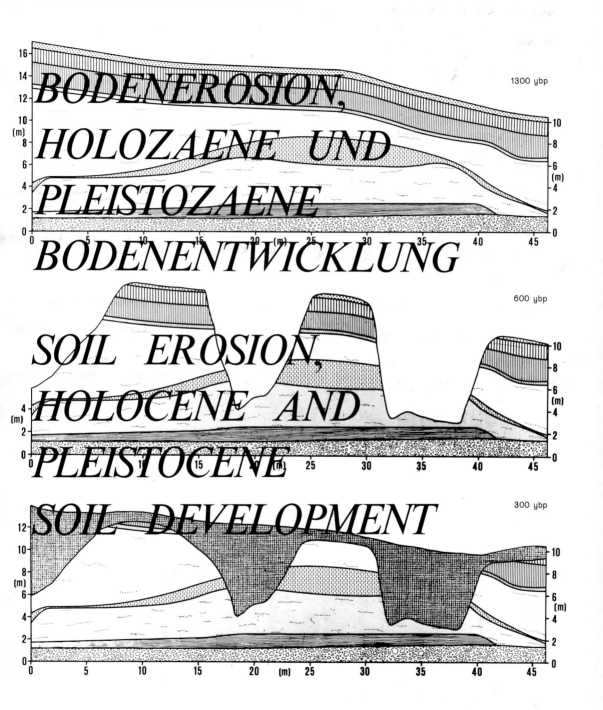

CATENA SUPPLEMENT 3

Jan de Ploey (Ed.)

RAINFALL SIMULATION, RUNOFF and SOIL EROSION

CATENA SUPPLEMENT 4, 1983

Price: DM 120,-

ISSN 0722-0723 ISBN 3-923381-03-4

This CATENA-Supplement may be an illustration of present-day efforts made by geomorphologists to promote soil erosion studies by refined methods and new conceptual approaches. On one side it is clear that we still need much more information about erosion systems which are characteristic for specific geographical areas and ecological units. With respect to this objective the reader will find in this volume an important contribution to the knowledge of active soil erosion, especially in typical sites in the Mediterranean belt, where soil degradation is very acute. On the other hand a set of papers is presented which enlighten the important role of laboratory research in the fundamental parametric investigation of processes, i.e. erosion by rain. This is in line with the progressing integration of field and laboratory studies, which is stimulated by more frequent feed-back operations. Finally we want to draw attention to the work of a restricted number of authors who are engaged in the difficult elaboration of pure theoretical models which may pollinate empirical research, by providing new concepts to be tested. Therefore, the fairly extensive publication of two papers by CULLING on soil creep mechanisms, whereby the basic force-resistance problem of erosion is discussed at the level of the individual particles.

All the other contributions are focused mainly on the processes of erosion by rain. The use of rainfall simulators is very common nowadays. But investigators are not always able to produce full fall velocity of waterdrops. EPEMA & RIEZEBOS give complementary information on the erosivity of simulators with restricted fall heights. MOEYERSONS discusses splash erosion under oblique rain, produced with his newly-built S.T.O.R.M.-1 simulator This important contribution may stimulate further investigations on the nearly unknown effects of oblique rain. BRYAN & DE PLOEY examined the comparability of erodibility measurements in two laboratories with different experimental set-ups They obtained a similar gross ranking of Canadian and Belgian topsoils

Both saturation overland flow and subsurface flow are important runoff sources u the rainforests of northeastern Queensland. Interesting, there, is the correlation between colour and hydraulic conductivity observed by BONELL, GILMOUR & CASSELLS. Ru generation was also a main topic of IMESON's research in northern Morocco, stressing mechanisms of surface crusting on clayish topsoils.

For southeastern Spain THORNES & GILMAN discuss the applicability of ero models based on fairly simple equations of the "Musgrave-type" After Richter (Germ and Vogt (France) it is TROPEANO who completes the image of erosion hazards in Euro vineyards. He shows that denudation is at the minimum in old vineyards, cultivated manual tools only. Also in Italy VAN ASCH collected important data about splash erosion rainwash on Calabrian soils. He points out a fundamental distinction between trans limited and detachment-limited erosion rates on cultivated fields and fallow land. F representative first order catchment in Central–Java VAN DER LINDEN comments trasting denudation rates derived from erosion plot data and river load measurements. too, on some slopes, detachment-limited erosion seems to occur

The effects of oblique rain, time-dependent phenomena such as crusting and ru generation, detachment-limited and transport-limited erosion including colluvial depos are all aspects of single rainstorms and short rainy periods for which particular, predi models have to be built. Moreover, it is argued that flume experiments may be an econ way to establish gross erodibility classifications. The present volume may give an impet further investigations and to the evaluation of the proposed conclusions and suggestio

Jan de Ploey

G.F. EPEMA & H.Th. RIEZEBOS
   FALL VELOCITY OF WATERDROPS AT DIFFERENT HEIGHTS AS A FACTOR INFLUENCING EROSIVITY OF SIMULATED RAIN

J. MOEYERSONS
   MEASUREMENTS OF SPLASH–SALTATION FLUXES UNDER OBLIQUE RAIN

R.B. BRYAN & J. DE PLOEY
   COMPARABILITY OF SOIL EROSION MEASUREMENTS WITH DIFFERENT LABORATORY RAINFALL SIMULATORS

M. BONELL, D.A. GILMOUR & D.S. CASSELLS
   A PRELIMINARY SURVEY OF THE HYDRAULIC PROPERTIES OF RAINFOREST SOILS IN TROPICAL NORTH–EAST QUEENSLAND AND THEIR IMPLICATIONS FOR THE RUNOFF PROCESS

A.C. IMESON
   STUDIES OF EROSION THRESHOLDS IN SEMI–ARID AREAS: FIELD MEASUREMENTS OF SOIL LOSS AND INFILTRATION IN NORTHERN MOROCCO

J.B. THORNES & A. GILMAN
   POTENTIAL AND ACTUAL EROSION AROUND ARCHAEOLOGICAL SITES IN SOUTH EAST SPAIN

D TROPEANO
   SOIL EROSION ON VINEYARDS IN THE TERTIARY PIEDMONTESE BASIN (NORTHWESTERN ITALY): STUDIES ON EXPERIMENTAL AREAS

TH.W.J. VAN ASCH
   WATER EROSION ON SLOPES IN SOME LAND UNITS IN A MEDITERRANEAN AREA

P VAN DER LINDEN
   SOIL EROSION IN CENTRAL–JAVA (INDONESIA). A COMPARATIVE STUDY OF EROSION RATES OBTAINED BY EROSION PLOTS AND CATCHMENT DISCHARGES

W.E.H. CULLING
   SLOW PARTICULARATE FLOW IN CONDENSED MEDIA AS AN ESCAPE MECHANISM: I. MEAN TRANSLATION DISTANCE

W.E.H. CULLING
   RATE PROCESS THEORY OF GEOMORPHIC SOIL CREEP

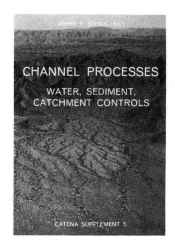

Asher P. Schick (Ed.):

CHANNEL PROCESSES
WATER, SEDIMENT, CATCHMENT CONTROLS

CATENA SUPPLEMENT 5, 1984

Price DM 110,—

ISSN 0722-0723 / ISBN 3-923381-04-2

PREFACE

Two decades ago, the publication of 'Fluvial Processes in Geomorphology' brought to maturity a new field in the earth sciences. This field – deeply rooted in geophy and geology and incorporating many aspects of hydrology, climatology, pedology – is well served by the forum provided by CATENA. Much progress has been accomplished in fluvial geomorphology during those twenty years, but highly complex and delicate relationships between channel processes and catchment controls still raise intriguing problems. Concepts dealing with threshold and systems, and modern tools such as remote sensing and sophisticated models, have not decisively resolved the simple but elusive dual problem: how does the catchment shape the stream channel and valley to its form, and why? How does the channel transmit its influence upstream in order to make the catchment what it is?

Partial solutions, in a regional or thematic sense, are common and important. Addition to contributing a building block to the study of fluvial geomorphology, they also produce a number of new questions. The consequent proliferation of such topics characterises this collection of papers. The basic tool of geomorphical interpretation – the magnitude, frequency, and mechanism of sediment water conveyance – is a prime focus of interest. Increasingly important in this context in recent years is the role of human interference natural fluviomorphic systems. Effects of drainage ditching, transport of pesticides absorbed in sediment, and the flushing of nutrients are some of the Manconditioned items mentioned in this volume. Other contributions deal with the intricate issue, especially in extreme climatic zones, between physical process generalization and macroregional morphoclimatic influences.

The contributions of PICKUP and of PICKUP & WARNER represent two of very few detailed quantitative geomorphological analyses of very humid tropical catchments. The 8 to 10 m mean annual rainfall in the equatorial mountain studied combines with effective landsliding to produce extremely high denudation rates. However, many aspects of channel behaviour are similar to those of arid rivers. Particularly interesting are the relationships derived between channel characteristics, perimeter sediment and bedload transport.

Several small ephemeral and intermittent streams in Ohio studied by THARP, although variable in catchment area and in peak discharge, have a similar competence; while sorting increases downstream, the coarsest sizes tend to remain constant. Sorting of fluvial sediment, though on a much longer time scale and in an arid climate, also plays an important role in the contribution by MAYER, GERSON & BULL. They find that modern channel sediment size exhibits the most rapid downstream decrease in mean particle size, while Pleistocene deposits show the least rapid decrease and are consistently finer than younger deposits. The difference is attributed to climatic change and a predictive model thereto is presented.

HASSAN, SCHICK & LARONNE describe a new method for the magnetic tracing of large bedload particles capable of detecting tagged particles redeposited by floods up to several decimetres below the channel bed surface. Their method may considerably enhance the value of numerous experiments with painted pebbles, previously reported or currently in progress. Suspended sediment is the subject of the paper by CARLING. He experiments with sampling gravel-bedded, flashy streams by two methods, and concludes that pump-sampling and 'bucket' sampling show significant differences only for very high discharges. Suspended sediment concentration is also dealt with by GURNELL & FENN, but in a proglacial environment – a climatic zone about which our knowledge is largely deficient. They find some correspondence between 'englacial' and 'subglacial' flow components and the total suspended sediment concentration.

The effects of human interference by ditching in a forest catchment on sediment concentration and sediment yield is discussed by BURT, DONOHOE & VANN. A local reservoir afforded an opportunity to monitor in detail the influence of these drainage operations on the sediment concentration which increased dramatically and, after several months, gradually recovered due to revegetation. TERNAN & MURGATROYD analyse sediment concentrations and specific conductance in a humid, forest and marsh environment. Permanent vegetation dams are found to influence sediment concentration directly through filtration and indirectly through changes in water depth and velocity. Changes in specific conductance are influenced by marsh inputs as well as by the addition of areas of coniferous forest. The relationship between quality of water and fluvial sediment characteristics is dealt with by HERRMANN, THOMAS & HÜBNER, who analyse the regional pattern of estuarine transport processes. They conclude that high pesticide concentrations are correlated with high concentrations of suspended sediment. Hydrodynamic rather than physicochemical factors influence the regional distribution in the estuary, and the effect of brooklets draining intensively cultivated land is quite evident.

Asher P. Schick

CONTENTS

PICKUP
GEOMORPHOLOGY OF TROPICAL RIVERS
I. LANDFORMS, HYDROLOGY AND SEDIMENTATION IN THE FLY AND LOWER PURARI, PAPUA NEW GUINEA

PICKUP & R. F. WARNER
GEOMORPHOLOGY OF TROPICAL RIVERS
II. CHANNEL ADJUSTMENT TO SEDIMENT LOAD AND DISCHARGE IN THE FLY AND LOWER PURARI, PAPUA NEW GUINEA

CARLING
COMPARISON OF SUSPENDED SEDIMENT RATING CURVES OBTAINED USING TWO SAMPLING METHODS

TERNAN & A. L. MURGATROYD
THE ROLE OF VEGETATION IN BASEFLOW SUSPENDED SEDIMENT AND SPECIFIC CONDUCTANCE IN GRANITE CATCHMENTS, S. W. ENGLAND

BURT, M. A. DONOHOE & A. R. VANN
CHANGES IN THE SEDIMENT YIELD OF A SMALL UPLAND CATCHMENT FOLLOWING A PRE-AFFORESTATION DITCHING

HERRMANN, W. THOMAS & D. HÜBNER
ESTUARINE TRANSPORT PROCESSES OF POLYCHLORINATED BIPHENYLS AND ORGANOCHLORINE PESTICIDES – EXE ESTUARY, DEVON

W. SEILER
MORPHODYNAMISCHE PROZESSE IN ZWEI KLEINEN EINZUGSGEBIETEN IM OBERLAUF DER ERGOLZ – AUSGELÖST DURCH DEN STARKREGEN VOM 29. JULI 1980

A. M. GURNELL & C. R. FENN
FLOW SEPARATION, SEDIMENT SOURCE AREAS AND SUSPENDED SEDIMENT TRANSPORT IN A PRO-GLACIAL STREAM

T. M. THARP
SEDIMENT CHARACTERISTICS AND STREAM COMPETENCE IN EPHEMERAL AND INTERMITTENT STREAMS, FAIRBORN, OHIO

L. MAYER, R. GERSON & W. B. BULL
ALLUVIAL GRAVEL PRODUCTION AND DEPOSITION – A USEFUL INDICATOR OF QUATERNARY CLIMATIC CHANGES IN DESERTS (A CASE STUDY IN SOUTHWESTERN ARIZONA)

M. HASSAN, A. P. SCHICK & J. B. LARONNE
THE RECOVERY OF FLOOD-DISPERSED COARSE SEDIMENT PARTICLES – A THREE-DIMENSIONAL MAGNETIC TRACING METHOD

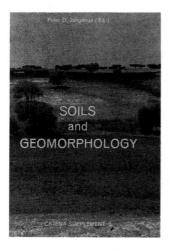

Peter D. Jungerius (Ed.):

Soils and Geomorphology

CATENA SUPPLEMENT 6 (1985)

Price DM 120,-

ISSN 0722-0723 / ISBN 3-923381-05-0

It was 12 years ago that CATENA's first issue was published with its ambitious subtitle "Interdisciplinary Journal of Geomorphology – Hydrology – Pedology". Out of the nearly one hundred papers that have been published in the regular issues since then, one-third have been concerned with subjects of a combined geomorphological and pedological nature. Last year it was decided to devote SUPPLEMENT 6 to the integration of these two disciplines. Apart from assembling a number of papers which are representative of the integrated approach, I have taken the opportunity to evaluate the character of the integration in an introductory paper. I have not attempted to cover the whole bibliography on the subject: an on-line consultation of the Georef files carried out on 29th October, 1984, produced 3627 titles under the combined keywords 'geomorphology' and 'soils'. Rather, I have made use of the ample material published in CATENA to emphasize certain points.

In spite of the fact that land forms as well as soils are largely formed by the same environmental factors, geomorphology and pedology have different roots and have developed along different lines. Papers which truly emanate the two lines of thinking are therefore relatively rare. This is regrettable because grafting the methodology of the one discipline onto research topics of the other often adds a new dimension to the framework in which the research is carried out. It is the aim of this SUPPLEMENT to stimulate the cross-fertilization of the two disciplines.

The papers are grouped into 5 categories: 1) the response of soil to erosion processes, 2) soils and slope development, 3) soils and land forms, 4) the age of soils and land forms, and 5) weathering (including karst).

P.D. Jungerius

P.D. JUNGERIUS
  SOILS AND GEOMORPHOLOGY

The response of soil to erosion processes

C.H. QUANSAH
  THE EFFECT OF SOIL TYPE, SLOPE, FLOWRATE AND THEIR INTERACTIONS ON DETACHMENT BY OVERLAND FLOW WITH AND WITHOUT RAIN

D.L. JOHNSON
  SOIL THICKNESS PROCESSES

Soils and slope development

M. WIEDER, A. YAIR & A. ARZI
  CATENARY SOIL RELATIONSHIPS ON ARID HILLSLOPES

D.C. MARRON
  COLLUVIUM IN BEDROCK HOLLOWS ON STEEP SLOPES, REDWOOD CREEK DRAINAGE BASIN, NORTHWESTERN CALIFORNIA

Soil and landforms

D.J. BRIGGS & E.K. SHISHIRA
  SOIL VARIABILITY IN GEOMORPHOLOGICALLY DEFINED SURVEY UNITS IN THE ALBUDEITE AREA OF MURCIA PROVINCE, SPAIN

C.B. CRAMPTON
  COMPACTED SOIL HORIZONS IN WESTERN CANADA

The age of soils and landforms

D.C. VAN DIJK
  SOIL GEOMORPHIC HISTORY OF THE TARA CLAY PLAINS S.E. QUEENSLAND

H. WIECHMANN & H. ZEPP
  ZUR MORPHOGENETISCHEN BEDEUTUNG DER GRAULEHME IN DER NORDEIFEL

M.J. GUCCIONE
  QUANTITATIVE ESTIMATES OF CLAY-MINERAL ALTERATION IN A SOIL CHRONOSEQUENCE IN MISSOURI, U.S.A.

Weathering (including Karst)

A.W. MANN & C.D. OLLIER
  CHEMICAL DIFFUSION AND FERRICRETE FORMATION

M. GAIFFE & S. BRUCKERT
  ANALYSE DES TRANSPORTS DE MATIERES ET DES PROCESSUS PEDOGENETIQUES IMPLIQUES DANS LES CHAINES DE SOLS DU KARST JURASSIEN

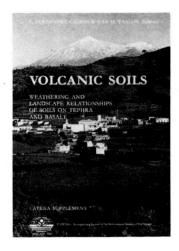

E. Fernandez Caldas & Dan H. Yaalon (Editors):

VOLCANIC SOILS
Weathering and Landscape
Relationships of Soils on Tephra and Basalt

CATENA SUPPLEMENT 7, 1985

Price DM 128,—

ISSN 0722-0723 / ISBN 3-923381-06-9

PREFACE

This CATENA SUPPLEMENT contains selected papers presented at the International meeting on Volcanic Soils held in Tenerife, July 1984. The meeting brought together over 80 scientists from 21 countries, with interest in the origin, nature and properties of soils on tephra and basaltic parent materials and their management. Some 51 invited and contributed papers and 8 posters were presented on a wide range of subjects related to volcanic soils, many of them dealing with weathering and landscape relationships. Classification was also discussed extensively during a six day excursion of the islands of La Palma, Gomera and Lanzarote, which enabled the participants to see the most representative volcanic soils of the Canary Archipelago under a considerable range of climatic regimes and parent material.

Because volcanic soils are not a common occurrence in regions where pedology developed and progressed during its early stages, recognition of their specific properties made an impact only in the late forties. The name **Ando** soils, now recognized as a special Great Group in all comprehensive soil classification systems, was coined in 1947 during reconnaissance soil surveys in Japan made by American soil scientists. Subsequently a Meeting on the Classification and Correlation of Soils from Volcanic Ash, sponsored by FAO and UNESCO, was held in Tokyo, Japan, in 1964, in preparation for the Soil Map of the World. This was followed by meetings of a Panel on Volcanic Ash Soils in Latin America, Turrialba, Costa Rica, in 1969 and a second meeting in Pasto, Colombia in 1972. At the International Conference on Soils with Variable Charge, Palmerston, New Zealand, 1981, the subject of Andisols was discussed intensively. Most recently the definitions of Andepts, as presented in the 1975 U.S. Soil Taxonomy, prompted the establishment of an International Committee on the Classification of Andisols (ICOMAND), chaired by M. Leamy from C.S.I.R., New Zealand, which held a number of international classification workshops, the latest in Chile and Ecuador, in January 1984. The continuous efforts to improve and revise the new classification of these soils is also reflected in some of the papers in this volume.

While Andosols or Andisols formed on tephra (volcanic ash), essentially characterized by low bulk density (less than 0.9 g/cm^3) and a surface complex dominated by active Al, cover worldwide an area of about 100 million hectares (0.8% of the total land area), the vast basaltic plateaus and their associated soils cover worldwide an even greater area, frequently with complex age and landscape relationships. While these soils do not generally belong to the ando group, their pedogenetic pathways are also strongly influenced by the nature and physical properties of the basalt rock. The papers in this volume cannot cover the wide variety of properties of the soils in all these areas, some of which have been reviewed at previous meetings. In this volume there is a certain emphasis on some of the less frequently studied environments and on methods of study and characterization as a means to advance the recognition and classification of these soils.

The Tenerife meeting was sponsored by a number of national and international organizations, including the Autonomous Government of the Canary Islands, the Institute of Ibero American Cooperation in Madrid, the Directorate on Scientific Policy of the Ministry of Education and Science, Madrid, the International Soil Science Society, ORSTOM of France, and ICOMAND. Members and staff of the Department of Soil Science of the University of La Laguna had the actual task of organizing the meeting and the field trips. In editing the book we benefitted from the manuscript reviews by many of our colleagues all over the world, and the capable handling and sponsorship of the CATENA VERLAG. To all those who have extended their help we wish to express warm thanks.

La Laguna and Jerusalem,
Summer 1984

E. Fernandez Caldas
D.H. Yaalon
Editors

CONTENTS

PARFITT & A.D. WILSON
ESTIMATION OF ALLOPHANE AND HALLOYSITE IN THREE SEQUENCES OF VOLCANIC SOILS, NEW ZEALAND

J. HERNANDEZ MORENO, V. CUBAS GARCIA, A. GONZALEZ BATISTA & E. FERNANDEZ CALDAS
STUDY OF AMMONIUM OXALATE REACTIVITY AT pH 6.3 (Ro) IN DIFFERENT TYPES OF SOILS WITH VARIABLE CHARGE. I

E. FERNANDEZ CALDAS, J. HERNANDEZ MORENO, M. TEJEDOR SALGUERO, A. GONZALEZ BATISTA & V. CUBAS GARCIA
BEHAVIOUR OF OXALATE REACTIVITY (Ro) IN DIFFERENT TYPES OF ANDISOLS. II

RADCLIFFE & G.P. GILLMAN
SURFACE CHARGE CHARACTERISTICS OF VOLCANIC ASH SOILS FROM THE SOUTHERN HIGHLANDS OF PAPUA NEW GUINEA

GONZALEZ BONMATI, M.P. VERA GOMEZ & J.E. GARCIA HERNANDEZ
KINETIC STUDY OF THE EXPERIMENTAL WEATHERING OF LOGITE AT DIFFERENT TEMPERATURES

JEZEBOS
HIGH-CONCENTRATION LEVELS OF HEAVY MINERALS IN TWO VOLCANIC SOILS FROM COLOMBIA:
A POSSIBLE PALEOENVIRONMENTAL INTERPRETATION

L.J. EVANS & W. CHESWORTH
THE WEATHERING OF BASALT IN AN ARCTIC ENVIRONMENT

R. JAHN, Th. GUDMUNDSSON & K. STAHR
CARBONATISATION AS A SOIL FORMING PROCESS ON SOILS FROM BASIC PYROCLASTIC FALL DEPOSITS ON THE ISLAND OF LANZAROTE, SPAIN

P. QUANTIN
CHARACTERISTICS OF THE VANUATU ANDOSOLS

P. QUANTIN, B. DABIN, A. BOULEAU, L. LULLI & D. BIDINI
CHARACTERISTICS AND GENESIS OF TWO ANDOSOLS IN CENTRAL ITALY

A. LIMBIRD
GENESIS OF SOILS AFFECTED BY DISCRETE VOLCANIC ASH INCLUSIONS, ALBERTA, CANADA

M.L. TEJEDOR SALGUERO, C. JIMENEZ MENDOZA, A. RODRIGUEZ RODRIGUEZ & E. FERNANDEZ CALDAS
POLYGENESIS ON DEEPLY WEATHERED PLIOCENE BASALT, GOMERA (CANARY ISLANDS): FROM FERRALLITIZATION TO SALINIZATION

M. Pécsi(Editor)

LOESS AND ENVIRONMENT

SPECIAL ISSUE ON THE OCCASION OF THE XII th International Congress of the INTERNATIONAL UNION OF QUATERNARY RESEARCH (INQUA) Ottawa 1987

CATENA SUPPLEMENT 9

160 pages / hardcover / price DM 128,-
Special rate for subscriptions until
December 15, 1987: DM 102,40

Date of publication: July 15, 1987   ORDER NO. 499/00108

ISSN 0722-0723/ISBN 3-923381-08-5

ORDER FORM

to: CATENA VERLAG,
    D-3302 CREMLINGEN 4, West Germany

ORDER NO. 499/00108
☐ Please, send me ..... copy(ies) at the special subscription rate at DM 119,20 (20% reduction)

☐ Please, enter a standing order for CATENA SUPPLEMENTS, starting with Supplement ..... (30%) reduction on the normal price)

☐ airspeeded plus DM 10.- (USA/CANADA), DM 15,- (JAPAN/BRAZIL), DM 20.- (Australia/New Zealand)

☐ cheque enclosed   ☐ send invoice

Name _____

ADDRESS _____

_____

Date/Signature _____

N E W

CATENA paperback

Joerg Richter
THE SOIL AS A REACTOR
Modelling Processes in the Soil

If we are to solve the pressing economic and ecological problems in agriculture, horticulture and forestry, and also with "waste" land and industrial emmissions, we must understand the processes that are going on in the soil. Ideally, we should be able to treat these processes quantitatively, using the same methods the civil engineer needs to get the optimum yield out of his plant. However, it seems very questionable, whether we would use our soils properly by trying to obtain the highest profit through maximum yield. It is vital to remember that soils are vulnerable or even destructible although or even because our western industrialised agriculture produces much more food on a smaller area than some ten years ago.

This book is primarily oriented on methodology. Starting with the phenomena of the different components of the soils, it describes their physical parameter functions and the mathematical models for transport and transformation processes in the soil. To treat the processes operationally, simple simulation models for practical applications are included in each chapter.

After dealing in the principal sections of each chapter with heat conduction and the soil regimes of material components like gases, water and ions, simple models of the behaviour of nutrients, herbicides and heavy metals in the soil are presented. These show how modelling may help to solve problems of environmental protection. In the concluding chapter, the problem of modelling salt transport in heterogeneous soils is discussed.

The book is intended for all scientists and students who are interested in applied soil science, especially in using soils effectively and carefully for growing plants: applied pedologists, land reclamation and improvement specialists, ecologists and environmentalists, agriculturalists, horticulturists, foresters, biologists (especially microbiologists), landscape planers and all kinds of geoscientists.

Prof. Dr. Joerg Richter
Institute of Soil Science
University of Hannover, FRG

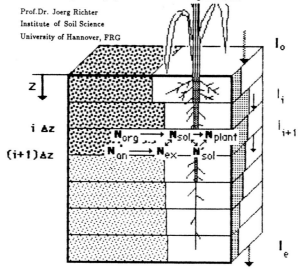

ISBN 3-923381-09-3   Price: DM 38.50

CATENA

AN INTERDISCIPLINARY JOURNAL OF
SOIL SCIENCE – HYDROLOGY – GEOMORPHOLOGY
FOCUSING ON
GEOECOLOGY AND LANDSCAPE EVOLUTION

EDITORS:

F. Ahnert, Aachen
G. Bartels, Koeln
L. Berry, Worcester, Mass.
J. J. Bigarella, Curitiba
H.-P. Blume, Kiel
J. Bouma, Wageningen
B. Bunting, Hamilton, Ont.
P. A. Burrough, Utrecht
K. Butzer, Austin
R. B. Bryan, Toronto
G. Castany, Orléans
J. Demek, Brno

I. Douglas, Manchester
G. H. Dury, Cambridge
A. R. Eschner, Syracuse, N.Y.
H. Faure, Marseille
H. Foelster, Goettingen
O. Fraenzle, Kiel
R. Herrmann, Bayreuth
K. Hirakawa, Yamanashi
J. v. Hoyningen-Huene, Braunschweig
A. C. Imeson, Amsterdam
P. D. Jungerius, Amsterdam
J. Kukla, Palisades
L. B. Leopold, Berkeley
J. A. Mabbutt, Kensington

B. Messerli, Bern
B. Meyer, Goettingen
P. Michel, Strasbourg
R. B. Morrison, Denver
D. Newson, Wallingford
R. Paepe, Bruessel
M. Pécsi, Budapest
J. de Ploey, Leuven
R. Pullan, Liverpool
R. V. Ruhe, Bloomington, Ind.
A. Ruellan, Paris
M. Sarnthein, Kiel
A. P. Schick, Jerusalem
A. Semmel, Frankfurt

O. Slaymaker, Vancouver
L. Starkel, Krakow
O. Strebel, Hannover
U. Streit, Muenster
J. B. Thornes, London
A. Velichko, Moskva
J. Vogt, Strasbourg
R. Webster, Harpenden
D. H. Yaalon, Jerusalem
A. Young, Norwich

CHIEF EDITOR:
H. Rohdenburg, Braunschweig

A Cooperating Journal of the International Society of Soil Science

CATENA publishes original contributions in the fields of

GEOECOLOGY, the geoscientific–hydro–climatological subset of process-oriented studies of the present ecosystem,
– the total environment of landscapes and sites
– the flux of energy and matter (water, solutes, suspended matter, bed load) with special regard to space-time variability
– the changes in the present ecosystem, including the earth surface,
and

LANDSCAPE EVOLUTION, the genesis of the present ecosystem, in particular the genesis of its structure concerning soils, sediment, relief, their spatial organization and analysis in terms of paleo-processes;
– soils: surface, relief and fossil soils, their spatial organization pertaining to relief development,
– sediment with relevance to landscape evolution, the paleohydrologic environment with respect to surface runoff, competence, and capacity for transport of bed material and suspended matter, infiltration, groundwater and channel flow,
– the earth's surface, relief elements and their spatial–hierarchical organization in relation to soils and sediment
– the paleoclimatological properties of the sequence of paleoenvironments

CATENA publishes multidisciplinary studies as well as monodisciplinary papers that are of interest to other disciplines and are of relevance to landscape studies.

ORDER FORM

To: SUBSCRIPTION DEPARTMENT **CATENA**, Brockenblick 8, 3302 Cremlingen 4, W. Germany
or your Subscription Agency

Please, enter a subscription starting with:

		Europe	Overseas
copy(ies) CATENA vol. 1, 1973/74, Vol. 1 No. 1/2 is expired	No. 3/4	DM 42.00	US $ 20.00
copy(ies) CATENA vol. 2, 1975,	No. 1-4	DM 79.00	US $ 35.90
copy(ies) CATENA vol. 3, 1976/77,	No. 1-4	DM 84.00	US $ 38.20
copy(ies) CATENA vol. 4, 1977,	No. 1-4	DM 96.00	US $ 43.50
copy(ies) CATENA vol. 5, 1978,	No. 1-4	DM 96.00	US $ 43.50
copy(ies) CATENA vol. 6, 1979,	No. 1-4	DM 107.50	US $ 61.43
copy(ies) CATENA vol. 7, 1980,	No. 1-4	DM 118.00	US $ 67.50
copy(ies) CATENA vol. 8, 1981,	No. 1-4	DM 135.00	US $ 78.00
copy(ies) CATENA vol. 9, 1982,	No. 1-4	DM 154.00	US $ 89.00
copy(ies) CATENA vol. 10, 1983,	No. 1-4	DM 173.00	US $ 98.00
copy(ies) CATENA vol. 11, 1984,	No. 1-4	DM 194.00	US $ 110.00
copy(ies) CATENA vol. 12, 1985,	No. 1-4	DM 219.00	US $ 105.00
copy(ies) CATENA vol. 13, 1986,	No. 1-4	DM 248.00	US $ 110.00
copy(ies) CATENA vol. 14, 1987	No. 1-6	DM 379.00	US$ 223.00

☐ air speeded each volume: U.S.A. plus US $ 10.–;
Japan, Brazil etc. US $ 15.–; Australia, New Zealand US $ 20.-

The prepaid **personal subscription with 25% reduction** is available from the publisher only. 50% reduction starting with vol. 14, 1987.
Unesco coupons accepted.

Date _____ Signature _____

Name _____ ☐ cheque enclosed

Address _____ ☐ send invoice

☐ send GUIDE FOR AUTHORS

SELECTED TITLES CATENA

SOIL DEVELOPMENT, SOIL CLASSIFICATION

Amit, Gerson: The Evolution of Holocene Reg (Gravelly) Soils in Deserts

Bork: Die holozäne Relief- und Bodenentwicklung in Lößgebieten - Beispiele aus dem südöstlichen Niedersachsen

Bouchard: Influences Stationelles sur l'Altération Chimique des Sols Dérivés de Till (Sherbrooke, Que., Canada)

Bowman: Iron Coating in Recent Terrace Sequences under Extremely Arid Conditions

Dan, Yaalon: Automorphic Saline Soils in Israel

Ellis: Micromorphological Aspects of Arctic-Alpine Pedogenesis in the Okstindan Mountains, Norway

Fränzle, Killisch, Ingenpass, Mich: Die Klassifizierung von Bodenprofilen als Grundlage agrarer Standortplanung in Entwicklungsländern: Ein Beispiel aus dem Savannengebiet Nordost-Ghanas

Gaucher: Vers une Classification Pédologique Naturelle Basée sur la Géochémie de la Pédogenèse

Gellatly: Phosphate Retention: Relative Dating of Holocene Soil Development

Hallsworth, Darley: Formation of Soils in an Aridic Environment Western New South Wales, Australia

Hoffmann, Blume: Holozäne Tonverlagerung als profilprägender Prozess lehmiger Landböden norddeutscher Jungmoränenlandschaften?

Johnson: Soil Thickness Processes

Kowalkowski, Brogowski: Features of Cryogenic Environment in Soils of Continental Tundra and Arid Steppe on the Southern Khangai Slope under Electron Microscope

Neumann, Blume: Struktur, Genese und Ökologie der Auenbodenschaft eines norddeutschen Fluß-Sees

Parker, Chartres: The Effect of Recent Land Use Changes on Red Podsolic Soils near Sydney, N.S.W., Australia

Remmelzwaal: Translocation and Transformation of Clay in Alfisols in Early, Middle and Late Pleistocene Coastal Sands of Southern Italy

Rohdenburg: Neue 14C-Daten aus Marokko und Spanien und ihre Aussagen für die Relief- und Bodenentwicklung im Holozän und Jungpleistozän

Summerfield: Distribution, Nature and Probable Genesis of Silcrete in Arid and Semi-Arid Southern Africa

Zaidenberg, Dan, Koyumdjisky: The Influence of Parent Material, Relief and Exposure on Soil Formation in the Arid Region of Eastern Samaria

WEATHERING, SOIL MINERALOGY, PARENT MATERIAL

Bateman, Catt: Modification of Heavy Mineral Assemblages in English Coversands by Acid Pedochemical Weathering
Bronger, Ensling, Kalk: Mineralverwitterung, Tonmineralneubildung und Rubefizierung in Terrae Calcis der Slowakei
Caine: Rock Weathering at the Soil Surface in an Alpine Environment
Calvo, Garcia-Rodeja, Macias: Mineralogical Variability in Weathering Microsystems of a Granite Outcrop of Galicia (Spain)
Dejou, Clement, de Kimpe: Importance du Site dans la Genèse des Minéraux Secondaires Issues des Altérations Superficielles
Dixon, Young: Character and Origin of Deep Arenaceous Weathering Mantles on the Bega Batholith, Southeastern Australia
Evans, Chesworth: The Weathering of Basalt in an Arctic Environment
Fedoroff, Goldberg: Comparative Micromorpholgy of Two Late Pleistocene Paleosols (in the Paris Basin)
Fedoroff, de Kimpe, Bourbeau: L'Altération des Minéraux Primaires en Milieu Podzolique en France Atlantique et au Québec
Fried: Äolische Komponenten in Rotlehmen des Adamaua-Hochlandes/Kamerun
González Bonmati, Vera Gómez, García Hernández: Kinetic Studies of the Experimental Weathering of Augite at Different Temperatures
Gudmundsson, Stahr: Mineralogical and Geochemical Alterations of "Podsol Bärhalde"
Kelletat: Studies on the Age of Honeycombs and Tafoni Features
Kowalkowski, Mycielska-Dowgiallo: Weathering of Quartz Grains in the Liquified Horizon of Permafrost Solonchaks in the Arid Steppe Zone, Central Mongolia
McGreevy, Smith: The Possible Role of Clay Minerals in Salt Weathering
Ollier: Weathering or Hydrothermal Alteration
Pullan: Termite Hills in Africa: Their Characteristics and Evolution
Pye: Granular Disintegration of Gneiss and Migmatites
Pye: Mineralogical and Textural Controls on the Weathering of Granitoid Rocks
Smalley, Krinsley: Loess Deposits Associated with Deserts
Smith, Whalley: Late Quaternary Drift Deposits of North Central Nigeria Examined by Scanning Electron Microscopy
Wilke, Duke, Jimoh: Mineralogy and Chemistry of Harmattan Dust in Northern Nigeria

SOIL GEOGRAPHY, SOIL TOPOSEQUENCES

Armstrong: Soils and Slopes in a Humid Temperate Environment: A Simulation Study
Blümel: Calcretes in Namibia, SE-Spain: Relations to Substratum, Soil Formation and Geomorphic Factors
Claridge, Campbell: A Comparison between Hot and Cold Desert Soils and Soil Processes
Dalsgaard, Baastrup, Bunting: The Influence of Topography on the Development of Alfisols on Calcareous Clayey Till in Denmark
Fehlberg, Stahr: Development of Sustained Land Use By Understanding Soil and Landscape Formation in the Desert Fringe Area of NW-Egypt
Fölster, von Christen: The Influence of Quaternary Uplift on the Altitude Zonation of Mountain Soils on Diabase and Volcanic Ash in Humid Parts of the Colombian Andes
Guthrie: Distribution of Great Groups of Aridisols in the United States
Kesel, Spicer: Geomorphologic Relationships and Ages of Soils on Alluvial Fans in the Rio General Valley, Costa Rica
Mabesoone, Lobo: Paleosols as Stratigraphic Indicators for the Cenozoic History of Northeastern Brazil
Reider: Geomorphic Implications of Pre-Wisconsin Soils on the White River Plateau Erosion Surface of Northwestern Colorado
Schwan, Cleveringa, Miedema: Pedogenic and Sedimentary Characteristics of a Late Glacial-Holocene Solifluction Deposit at Hjerupgyde, Denmark
Semmel: Catenen der feuchten Tropen und Fragen ihrer Geomorphologischen Deutung
Semmel: Böden und Relief im Grund- und Deckgebirge des Harer-Plateaus (Ost-Äthiopien)
Semmel, Rohdenburg: Untersuchungen zur Boden- und Reliefentwicklung in Süd-Brasilien
Soyer: Microrelief de Buttes Basses sur Sols Inondés Saisonnièrement au Sud-Shaba (Zaire)
Swanson: Soil Catenas on Pinedale and Bull Lake Moraines, Willow Lake, Wind River Mountains, Wyoming
Venzke: Bodentypen und Bodentypenvergesellschaftungen in Island
Wieder, Yair, Arzi: Catenary Soil Relationships on Arid Hillslopes
Xu, Webster: A Geostatistical Study of Topsoil Properties in Zhangwu County, China
van Dijk: Soil Geomorphic History of the Tara Clay Plains, S.E. Queensland

SOIL STRUCTURE, SOIL HYDROLOGY

Bariac, Klamecki, Jusserand, Létolle: Evolution de la Composition Isotopique De l'Eau (18 O) dans le Continuum Sol-Plante-Atmosphère

Becher, Vogel: Gefügemerkmale von Lößprofilen unterschiedlicher Pseudovergleyung in Südbayern

Blume, Vahrson, Meshref: Dynamics of Water, Temperature and Salts in Typical Aridic Sulfate Soils after Artificial Rainstorms

Bonell: Some Comments on the Association between Saturated Hydraulic Conductivity and Texture of Holderness Boulder Clay

Bonell, Gilmour, Cassells: A Preliminary Survey of the Hydraulic Properties of Rainforest Soils in Tropical North-East Queensland and their Implications for the Runoff Processes

Burghardt: Untersuchungen zur räumlichen und zeitlichen Variabilität der Wasserleitfähigkeit eines Flußmarsch-Bodens

Cousen, Farres: The Role of Moisture Content in the Stability of Soil Aggregates from a Temperate Silty Soil to Raindrop Impact

Farres: Some Observations on the Stability of Soil Aggregates to Raindrop Impact

Gabriels, de Boodt: Relationship between Moisture Content, Aggregate Formation and Aggregate Stability of a Loam Soil Treated with Soil Conditioners in Various Concentrations

Grieve: The Magnitude and Significance of Soil Structural Stability Declines under Cereal cropping

Grieve: A Comparison of Aggregate Stability Test Procedures in Determining the Stability of Fine Soils in Fife, Scotland

Gupta, Verma: Hydraulic Conductivity of a Swelling Clay in Relation to Irrigation Water Quality

Hartge: Böden als Teile von Systemen

Hartmann, Verplancke, de Boodt: Water Storage in a Bare and Cropped Sandy Loam

Horn, Knickrehm, Mattiat: Die Ermittlung der Porengrößenverteilung eines Tonbodens über Entwässerung und auf Rasterelektronenmiskroskopischem Wege - Ein Vergleich

Imeson, Vis, de Water: The Measurement of Water-Drop Impact Forces with a Piezo-Electric Transducer

Klotz: Das Verdünnungsverfahren zur Bestimmung der Durchlässigkeiten poröser Medien großer Körnung

Mbagwu: Estimating Dry-Season Crop Water Requirements from Climatological and Soil Available Water Capacity Data in the Sedimentary and Basement Complex Areas of Southern Nigeria

Rambal, Ibrahim, Rapp: Variabilité Spatiale des Variations du Stock d'Eau du Sols du Forêt - Application a l'Optimisation d'un Dispositif de Mesure du Bilan Hydrique

Ranade, Gupta: Salt Effect on Hydraulic Conductivity of a Vertisol

Wilson, Smart: Pipes and Pipe Flow Process in an Upland Catchment, Wales

SOIL CHEMISTRY

Archer, Maeckel: Calcrete Deposits on the Lusaka Plateau of Zambia
Balabane: Oxygen-18, Deuterium and Carbon-13 Content of Organic Matter from Litter and Humus Layers in Podzolic Soils
Conacher: Throughflow as a Mechanism Responsible for Excessive Soil Salinisation in None-Irrigated, Previously Arable Lands in the Western Australian Wheat Belt: A Field Study
Crowther: Carbon Dioxide Concentrations in Some Tropical Karst Soils, West Malaysia
Culling: Highly Erratic Spatial Variability of Soil-pH on Iping Common, West Sussex
Dormaar, Lutwick: Extractable Fe and Al as an Indicator for Buried Soil Horizons
Fallek, Ghiglione, Negre: Charactéristiques des Humoacides et des Humines des deux Rankers Pyrénéens
Gaiffe, Bruckert: Analyse des Transports de Matières et des Processus Pédogénetiques Impliqués dans les Chaines de Sols du Karst Jurassien
Grieve, Foster, Carter: Spatial and Temporal Variations in Concentrations of Three Ions in Solutions Extracted from a Woodland Soil
Gunn, Trudgill: Carbon Dioxide Production and Concentrations in the Soil Atmosphere: A Case Study from New Zealand Volcanic Ash Soils
James, Wharfe, Pegg, Clarke: A Cation Budget Analysis for a Coastal Dune System in North-West England
Leirós de la Peña, Guitián Ojea: A Comparison of Redox Processes in Coastal and Inland Hydromorphic Soils in Galicia (N.W. Spain)
Madsen, Jensen, Jakobsen, Platou: A Method for Identification and Mapping Potentially Acid Sulfate Soils in Jutland, Denmark
Mann, Ollier: Chemical Diffusion and Ferricrete Formation
Mariotti, Pierre, Vedy, Bruckert, Guillemot: The Abundance of Natural Nitrogen 15 in the Organic Matter of Soils along an Altitudinal Gradient (Chablais, Haute Savoie, France)
Mazzarino, Fölster: Freisetzung und Verteilung von Al- und Si-Oxiden in Mitteldeutschen Löß-Böden unter Wald
Nahon, Lappartient: Time Factor and Geochemistry in Iron Crust Genesis
Scharpenseel, Eichwald, Haupenthal, Neue: Zinc Deficiency in a Soil Toposequence, Grown to Rice, at Tiaong, Quezon Province, Philippines
Süßmann: Wasserqualität von Quell- und Dränwässern im Waldecker Buntsandsteingebiet
Vreeken: Variability of Depth to Carbonates in Fingertip Loess Watersheds in Iowa

BIOGEOCHEMISTRY, ENVIRONMENT

Drake: Groundwater Chemistry in the Schefferville, Quebec Iron Deposits

Foster: Chemistry of Bulk Precipitation Throughfall, Soil Water and Stream Water in a Small Catchment in Devon, England

Foster: Seasonal Solute Behaviour of Stormflow in a Small Agricultural Catchment

Francis, Thornes, Romero Diaz, Lopez Bermudez, Fisher: Topographic Control of Soil Moisture, Vegetation Cover and Land Degradation in a Moisture Stressed Mediterranean Environment

Frevert: A Statistical Model on the Fertility of Giant Taro Cultivation Pits at Tawara Atoll (Central Pacific)

Herrmann, Thomas, Hübner: Estuarine Transport Processes of Polychlorinated Biphenyls and Organochlorine Pesticides, Exe Estuary, Devon

Herrmann, Thomas, Schrimpff: Pollutant (Zn, PO_4, NO_2, PAH, BHC, Copostranol) Transport and its Modeling in a Small Stream

Herrmann: Regional Patterns of Polycyclic Aromatic Hydrocarbons in NE-Bavarian Snow and their Relationships to Anthropogenic Influence and Air Flow

Ho: Chemical Environment of Coastal Marshes and Swamps, Louisiana

James, Wharfe: The Chemistry of Rainwater in a Coastal Locality of Northwest England

Kienholz: Kombinierte Geomorphologische Gefahrenkarte 1:10000 von Grindelwald

Lichtfuß, Brümmer: Natürlicher Gehalt und anthropogene Anreicherung von Schwermetallen in den Sedimenten von Elbe, Eider, Trave und Schwentine

Reynolds, Hornung, Stevens: Solute Budgets and Denudation Rate Estimates for a Mid-Wales Catchment

Rump, Symader, Herrmann: Mathematical Modeling of Water Quality in Small Rivers (Nutrients, Pesticides and other Chemical Properties)

Symader: Suspended Heavy Metals - An Investigation of their Temporal Behaviour in Flowing Water

Symader, Thomas: Interpretation of Average Heavy Metal Pollution in Flowing Waters and Sediment by Means of Hierarchical Grouping Analysis Using Two Different Error Indices

van der Poel: Influence of Environmental Factors on the Growth of the Beech

Wolfenden, Lewin: Distribution of Metal Pollutants in Active Stream Sediments

Zeman: Hydrochemical Balance of a British Columbia Mountainous Watershed

HYDROLOGY

Ben-Zvi, Cohen: Frequency and Magnitude of Flows in the Negev
Bork, Rohdenburg: Transferable Parameterization Methods for Distributed Hydrological and Agroecological Catchment Models
Browne: Derivation of a Geological Index for Low Flow Studies
Day: Lithologic Controls of Drainage Density: A Study of Six Small Rural Catchments in New England, N.S.W.
de Ploey: A Stemflow Equation for Grasses and Similar Vegetation
Dury: Step-Functional Analysis of Long Records of Streamflow
Faber: River Discharge in Relation to Rainfall, Evapotranspiration and Lithology in the Serayu River Basin, Central Java, Indonesia
Grabczak, Maloszewski, Rozanski, Zuber: Estimation of the Tritium Input Function with the Aid of Stable Isotopes
Herrmann, Stichler: Groundwater-Runoff Relationships
Herrmann: Ein Anwendungsversuch der mehrdimensionalen Diskriminanzanalyse auf die Abflußvorhersage
Jackson: Inter-Station Rainfall Correlation under Tropical Conditions
Maloszewski, Zuber: Theoretical Possibilities of the 3H-3He Method in the Investigations of Groundwater Systems
McMahon: The Relief and Land Form Map of Australia: Does It Show Rock Types and Land Forms of Hydrologic Significance?
Rohdenburg, Diekkrüger, Bork: Deterministic Hydrological Site and Catchment Models for the Analysis of Agroecosystems
Smart, Wilson: Two Methods for the Tracing of Pipe Flow on Hillslopes
Srdoc, Obelic, Horvatincic, Sliepcevic, Stichler, Moser, Geyh: Isotope Analysis of Groundwater on the North African Plain
Victor, Sastry: Short Term Probabilities of Rainfall in a Semi-arid Monsoonal Climate
Woo: Hydrology of a Small Canadian High Arctic Basin During the Snowmelt Period

RIVERS, LAKES, CHANNEL PROCESSES

Bäumer: A Numerical Model Computing Distributions of Horizontal Components of Flow Velocity in Shallow, Tideless River Mouths - Case Study: Mouth of River Schussen, an Affluent to Lake Constance
Belperio: The Combined Use of Wash Load and Bed Load Rating Curves for the Calculation of Total Load: An Example from the Burdekin River, Australia

Ergenzinger, Conrady: A New Tracer Technique for Measuring Bedload in Natural Channels

Finlayson: Field Calibration of a Recording Turbidity Meter

Hassan, Schick, Laronne: The Recovery of Flood-Dispersed Coarse Sediment Particles - A Three-Dimensional Magnetic Tracing Method

Klein: Anti Clockwise Hysteresis in Suspended Sediment Concentration During Individual Storms: Holbeck Catchment; Yorkshire, England

Knighton: Longitudinal Changes in the Size and Shape of Stream Bed Material: Evidence of Variable Transport Conditions

Knighton: Channel Gradient in Relation to Discharge and Bed Material Characteristics

Lekach, Schick: Evidence for Transport of Bedload in Waves: Analysis of Fluvial Sediment Samples in a Small Upland Stream Channel

Mayer, Gerson, Bull: Alluvial Gravel Production and Deposition: A Useful Indicator of Quaternary Climatic Changes in Deserts

Merot, Bourguet, Le Leuch: Analyse d'une Crue a l'Aide du Traçage Naturel par l'Oxygene 18 Mesure dans les Pluies, le Sol, le Ruisseau

Pachur, Denner, Walter: A Freezing Device for Sampling the Sediment-Water Interfaces of Lakes

Petts, Pratts: Channel Changes Following Reservoir Construction on a Lowland English River

Petts, Thoms: Channel Aggradation Below Chew Valley Lake, Somerset, U.K.

Schick, Lekach: High Bedload Transport Rates in Relation to Stream Power, Wadi Mikeimin, Sinai

Tharp: Sediment Characteristics and Stream Competence in Ephemeral and Intermittent Streams, Fairborn, Ohio

Webb, Walling: The Magnitude and Frequency Characteristics of Fluvial Transport in a Devon Drainage Basin and some Geomorphological Implications

Zimmermann, Baumann, Imevbore, Henderson, Adeniji: Study of the Mixing Pattern of Lake Kainji Using Stable Isotopes

SOIL EROSION, SURFACE RUNOFF, SEDIMENTATION

Augustinus, Nieuwenhuyse: Soil Erosion in Vineyards in the Ardeche Region

Band: Field Parameterization of an Empirical Sheetwash Transport Equation

Bonell, Coventry, Holt: Erosion of Termite Mounds under Natural Rainfall in Semiarid Tropical Northeastern Australia

Bryan, Campbell: Sediment Entrainment and Transport During Local Rainstorms in the Steveville Badlands, Alberta

Bryan, de Ploey: Comparability of Soil Erosion Measurements with Different Laboratory Rainfall Simulators

Epema, Riezebos: Fall Velocity of Waterdrops at Different Heights as a Factor Influencing Erosivity of Simulated Rain

Farres: The Dynamics of Rainsplash Erosion and the Role of Soil Aggregate Stability

Govers: Selectivity and Transport Capacity of Thin Flows in Relation to Rill Erosion

Imeson: Studies of Erosion Thresholds in Semi-Arid Areas: Field Measurements of Soil Loss and Infiltration in Northern Morocco

Imeson, Verstraten: The Erodibility of Highly Calcareous Soil Material from Southern Spain

Klein: A Quantitatitive Approach to the Analysis of Slope Roughness and Effective Slope Angle

Logie: Influence of Roughness Elements and Soil Moisture on the Resistance of Sand to Wind Erosion

Luk, Morgan: Spatial Variations of Rainwash and Runoff within Apparently Homogenous Areas

Luk: Effect of Antecedent Soil Moisture Content on Rainwash Erosion

McGregor: An Investigation of Soil Erosion in the Colombian Rainforest Zone

Moeyersons: Measurements of Splash-Saltation Fluxes under Oblique Rain

Poesen, Savat: Detachment and Transportation of Loose Sediments by Raindrop Splash - Part II: Detachibility and Transportability Measurements

Savat, Poesen: Detachment and Transportation of Loose Sediments by Raindrop Splash - Part I: The Calculation of Absolute Data on Detachibility and Transportability

Savat: Common and Uncommon Selectivity in the Process of Fluid Transportation: Field Observations and Laboratory Experiments on Bare Surfaces

Seiler: Meßeinrichtungen zur Quantitativen Bestimmung des Geoökofaktors Bodenerosion in der Topologischen Dimension auf Ackerflächen im Schweizer Jura (südöstlich Basel)

van Asch: Water Erosion on Slopes in Some Land Units in a Mediterranean Area

van Hooff, Jungerius: Sediment Source and Storage in Small Watersheds on the Keuper Marls in Luxembourg, as Indicated by Soil Profile Truncation and the Deposition of Colluvium

van Zon: The Transport of Leaves and Sediment over a Forest Floor: A Case Study in the Grand Duchy of Luxembourg

Yair, de Ploey: Field Observations and Laboratory Experiments Concerning the Creep Process of Rock Blocks in an Arid Environment

LANDFORM EVOLUTION, SLOPE PROCESSES
(without soil erosion)

Ahnert: Untersuchungen über das Morphoklima und die Morphologie des Inselberggebietes von Machakos, Kenia

Bartels, Steinmann: Quartärgeomorphologische Untersuchungen im Nordteil der "Tunesischen Dorsale"

Brunotte: Quaternary Piedmont Plains on Weakly Resistant Rocks in the Lower Saxonian Mountains, (W. Germany)

Carter, Ciolkosz: Sorting and Thickness of Waste Mantle Material on a Sandstone Spur in Central Pennsylvania

Castleden: Fluvioperiglacial Pedimentation: A General Theory of Fluvial Development in Cool Temperate Lands, Illustrated from Western and Central Europe

Chester, Duncan: The Interaction of Volcanic Activity in Quaternary Times upon the Evolution of the Alcantara and Simeto Rivers, Mount Etna, Sicily

Culling: Slow Particularate Flow in Condensed Media as an Escape Mechanism: Mean Translocation Distance

Culling: Rate Process Theory of Geomorphic Soil Creep

de Ploey, Cruz: Landslides in the Serra do Mar, Brazil

East: A Multivariate Analysis of the Particle Size Characteristics of Regolith in a Catchment on the Darling Downs, Australia

Embrechts, de Dapper: Morphology and Genesis of Hillslope Pediments in the Febe Area (South Cameroon)

Lancaster: Formation of the Holocene Lake Chilwa Sand Bar, Southern Malawi

Lancaster: Dynamics of Deflation Hollows in the Elands Bay Area, Cape Province, South Africa

Lauritzen: Some Estimates of Denudation Rates in Karstic Areas of the Saltfjellet-Svartisen Region, North Norway

Modenesi: Weathering and Morphogenesis in a Tropical Plateau

Neuland: Diskriminanzanalytische Untersuchungen zur Identifikation der Auslösefaktoren für Rutschungen in verschiedenen Höhenstufen der Kolumbianischen Anden

Pain: Scarp Retreat and Slope Development near Picton, New South Wales, Australia

Pilgrim, Puvaneswaran, Conacher: Factors Affecting Natural Rates of Slope Development

Puvaneswaran, Conacher: Extrapolation of Short-Term Process Data to Long-Term Landform Development: A Case Study from Southwestern Australia

Rohdenburg: Geomorphologisch-Bodenstratigraphischer Vergleich zwischen dem nordostbrasilianischen Trockengebiet und immerfeucht-tropischen Gebieten Südbrasiliens mit Ausführungen zum Problemkreis der Pediplain-Pediment-Terrassentreppen

Rohdenburg, Sabelberg, Wagner: Sind konkave und konvexe Hänge prozeß-spezifische Formen?

van Asch: Landslides: The Deduction of Strength Parameters of Materials from Equilibrium Analysis

QUATERNARY, STRATIGRAPHY, CHRONOLOGY, PALEOSOLS, GEO-ARCHEOLOGY

Brook: Stratigraphic Evidence of Quaternary Climatic Change at Echo Cave, Transvaal, and a Paleoclimatic Record for Botswana and Northeastern South Africa
Butzer: Geomorphology and Geo-Archeology at Elandsbaai Western Cape, South Africa
Dan: Soil Chronosequences in Israel
Edwards, Thompson: Magnetic, Palynological and Radiocarbon Correlation and Dating Comparisons in Long Cores from a Northern Irish Lake
Felix-Henningsen, Urban: Paleoclimatic Interpretation of a Thick Intra-Saalian Paleosol - The "Bleeched Loam" on the Drenthe Moraines of Northern Germany
Finkl jun.: Stratigraphic Principles and Practices Related to Soil Mantles
Fitze: Zur Relativdatierung von Moränen aus der Sicht der Bodenentwicklung in den kristallinen Zentralalpen
Goldberg: Micromorphology of Sediments from Hayonim Cave, Israel
Heine: Radiocarbon Chronology of Late Quaternary Lakes in the Kalahari, Southern Africa
Hirakawa: Chronology and Evolution of Landforms During the Late Quaternary in the Tokachi Plain and Adjactent Areas, Hokkaido, Japan
Holliday: Early and Middle Holocene Soils at the Lubbock Lake Archeological Site, Texas
Macaire: Sequences of Polycyclic Soils Formed on Plio-Quaternary Alluvial Deposits in South-Western Paris Basin (France)
Mahaney, Boyer: Microflora Distributions in Quaternary Paleosols on Mount Kenya, East Africa
McFadden, Wells, Dohrenwend: Influences of Quaternary Climatic Changes on Processes of Soil Development on Desert Loess Deposits of the Cima Volcanic Field, California
Reider: Late Pleistocene and Holocene Soils of the Carter-McGee Archeological Site, Powder River Basin, Wyoming
Ricken: Mittel- und jungpleistozäne Lößdecken im südwestlichen Harzvorland - Stratigraphie, Paläopedologie, fazielle Differenzierung und Konnektierung in Flußterrassen
Sabelberg: The Stratigraphic Record of Late Quaternary Accumulation Series in South West Morocco and its Consequences Concerning the Pluvial Hypothesis
Schubert: Contribution to the Paleolimnology of Lake Valencia, Venezuela: Seismic Stratigraphy
Shaw, Cooke: Geomorphic Evidence for Late Quaternary Paleoclimates of the Middle Kalahari of Northern Botswana
Wyrwoll: The Sedimentology, Stratigraphy and Paleoenvironmental Significance of a Late Pleistocene Alluvial Fill: Central Coastal Areas of Western Australia
Zenses: Vergleichende Untersuchungen zum Bau pleistozäner Schuttablagerungen

ORDER FORM

to: Subscription Department CATENA
D-3302 Cremlingen–Destedt

Please enter a subscription for CATENA starting with Vol. 13, 1986

                US$ 146.-   DM 248.-

current volume: Vol. 14, 1987

ORDER NO. 499/04014   Price: DM 379.-
                                        US$ 223.-

50% reduction for personal subscribers starting with 1987

1986 personal subscribers receive 25% reduction

Advanced payment is required

☐ cheque enclosed   ☐ send invoice

Name: _____

Address: _____

_____

Date/Signature _____